SACRAMENTO PUBLIC LIBRARY

3 3029 04744 3932

C

SA   RAMENTO, CA 95814

D0484099

JUL   2002

BLUE GOLD

BLUE GOLD

*The Fight to Stop the Corporate Theft
of the World's Water*

MAUDE BARLOW
TONY CLARKE

The New Press
New York

© 2002 by Maude Barlow and Tony Clarke
All rights reserved.
No part of this book may be reproduced, in any form,
without written permission from the publisher.

Published in the United States by The New Press, New York, 2002
Distributed by W. W. Norton & Company, Inc., New York

Every reasonable effort has been made to contact the holders of
copyright for materials quoted in the book. The authors and
publisher will gladly receive information that will enable them
to rectify any inadvertent errors or omissions in subsequent editions.

ISBN 1-56584-731-8
CIP data available.

The New Press was established in 1990 as a not-for-profit alternative
to the large, commercial publishing houses currently dominating
the book publishing industry. The New Press operates in the public
interest rather than for private gain, and is committed to publishing,
in innovative ways, works of educational, cultural, and community
value that are often deemed insufficiently profitable.

The New Press, 450 West 41st Street, 6th floor, New York, NY 10036
www.thenewpress.com

Book design by Tannice Goddard

Printed in Canada

10 9 8 7 6 5 4 3 2 1

For KIMY PERNIA DOMICO,
tireless fighter for Indigenous rights to water,
who was "disappeared" by Colombian paramilitary forces
on June 2, 2001. You are dearly missed.

CONTENTS

ACKNOWLEDGMENTS

We are deeply indebted to an international family of water warriors, fighting to preserve water as a common heritage for all people and Nature. You are too many now, miraculously, to name here, but you are our inspiration and our joy. We especially wish to thank Jamie Dunn of the Blue Planet Project and The Council of Canadians for his tireless work in building an international water security movement, and Darren Puscas, of the Polaris Institute, for his excellent research on the transnational water corporations. Patricia Perdue, of The Council of Canadians, was, as always, cheerful, encouraging, generous, and gracious in her assistance to us. Don Bastian of Stoddart gave us tremendous moral support, and once again, we relied on the professional and ever thoughtful guidance of our editor, Kathryn Dean. We are also most grateful to our wonderful families for their understanding and support in helping bring this book to life.

The Council of Canadians' website which carries Blue Planet Project materials is www.canadians.org.

Maude Barlow and Tony Clarke
Ottawa, Canada, December 2001

INTRODUCTION

Watersheds come in families; nested levels of intimacy. On the grandest scale the hydrologic web is like all humanity — Serbs, Russians, Koyukon Indians, Amish, the billion lives in the People's Republic of China — it's broadly troubled, but it's hard to know how to help. As you work upstream toward home, you're more closely related. The big river is like your nation, a little out of hand. The lake is your cousin. The creek is your sister. The pond is her child. And, for better or worse, in sickness and in health, you're married to your sink.

— MICHAEL PARFIT, NATIONAL GEOGRAPHIC

Suddenly it is so clear: the world is running out of fresh water. Humanity is polluting, diverting, and depleting the wellspring of life at a startling rate. With every passing day, our demand for fresh water outpaces its availability and thousands more people are put at risk. Already, the social, political, and economic impacts of water scarcity are rapidly becoming a destabilizing force, with water-related conflicts springing up around the globe. Quite simply, unless we dramatically change our ways, between one-half and two-thirds of humanity will be living with severe fresh water shortages within the next quarter-century.

It seemed to sneak up on us. Until the last decade, the study of fresh water was left to highly specialized groups of experts — hydrologists, engineers, scientists, city planners, weather forecasters, and others with a niche interest in what so many of us took for granted. Now, however, an increasing number of voices — Worldwatch Institute, World Resources Institute, United Nations Environment Programme, International Rivers Network, Greenpeace, Clean Water Network, Sierra Club, and Friends of the Earth International, along with thousands of community groups around the world — are sounding the alarm: the global fresh water crisis looms as perhaps the greatest threat ever to the survival of our planet.

Tragically, this global call for action comes in an era guided by the principles of the so-called "Washington Consensus," a model of economics rooted in the belief that liberal market economics constitute the one and only economic choice for the whole world. Key to this "consensus" is the commodification of "the commons." Everything is for sale, even those areas of life, such as social services and natural resources, that were once considered the common heritage of humanity. Governments around the world are abdicating their responsibility to protect the natural resources within their borders, giving authority away to private companies that make a business of resource exploitation.

Faced with the now well-documented fresh water crisis, governments and international institutions are advocating a "Washington Consensus" solution: the privatization and commodification of water. Price water, they say in chorus; put it up for sale and let the market determine its future. For them, the debate is closed. Water, according to the World Bank and the United Nations, is a *human need*, not a *human right*. These are not semantics; the difference in interpretation is crucial. A human need can be supplied in many ways, especially for those with money. But no one can sell a human right.

When water was defined as a commodity at the second "World Water Forum" in The Hague in March 2000, government representatives at a parallel meeting did nothing to effectively counteract the statement. Instead, governments have helped pave the way for private corporations to sell water, for profit, to the thirsty citizens of the world. So a handful of transnational corporations, backed by the World Bank and the International Monetary Fund (IMF), are now aggressively taking over the management of public water services, dramatically raising the price of water to the local residents and profiting especially from the Third World's desperate search for solutions to its water crisis. Some are startlingly open about their motives; the decline in fresh water supplies and standards has created a wonderful venture opportunity for water corporations and their investors, they boast. The agenda is clear: water should be treated like any other tradable good, its use and distribution determined by the principles of profit.

At the same time, governments are signing away their control over domestic water supplies to trade agreements such as the North American Free Trade Agreement (NAFTA); its proposed successor, the Free Trade Area of the Americas (FTAA); and the World Trade Organization (WTO). These global trade institutions effectively give transnational corporations unprecedented access to the fresh water of signatory countries. Already, corporations have started to sue governments in order to gain access to domestic water sources, and armed with the protection of these international trade agreements, they are setting their sights on the mass transport of bulk water by diversion and by supertanker.

So far, most of this activity has taken place without public consultation or public input. The assumption has been made by the powerful forces of governments and the corporate sector that the debate is over: "everyone" agrees to the commodification of water. And yet no one has given the world's citizens a real opportunity to debate the hard political questions about water: Who owns it? Should anyone own it? If water is privatized, who will buy it for Nature? How will it be made available to the poor? Who gave transnational corporations the right to buy whole water systems? Who will protect water resources if they are taken over by the private sector? What is the role of government in the stewardship of water? How do those in water-rich countries share with those in water-poor countries? Who is the custodian of Nature's lifeblood? How do ordinary citizens become involved in the discussion?

This book presents some answers to these questions, answers based on a set of principles very different from those of the "Washington Consensus." We believe that fresh water belongs to the earth and all species and that no one has the right to appropriate it for personal profit. Water is part of the world's heritage and must be preserved in the public domain for all time and protected by strong local, national, and international law. At stake is the whole notion of "the commons," the idea that through our public institutions we recognize shared humanity and natural resources to be preserved for future generations.

We believe that access to clean water for basic needs is a fundamental human right; this vital resource cannot become a commodity sold to the highest bidder. Each generation must ensure that the abundance and quality of water is not diminished as a result of its activities. Great efforts must be made to restore the health of aquatic ecosystems that have already been degraded and to protect others from harm. Local and regional communities must be the watchdogs of our waterways and must establish principles that oversee the use of this precious resource.

Above all, we need to radically restructure our societies and lifestyles in order to reverse the drying of the earth's surface; we must learn to live within the watershed ecosystems that were created to sustain life. And we must abandon the specious notion that we can carelessly abuse the world's precious water sources because, somehow, technology will come to the rescue. There is no technological "fix" for a planet that has run out of water.

The debate over the wise and equitable use of the earth's water resources is far from over. In fact, it is just beginning. In this book, we tell the story of the world's growing fresh water crisis, the corporate assault on the water "commons," and the complicity of governments and international institutions in the theft of the world's fresh water. Most important, we show how ordinary citizens all over the world are engaging in a new form of citizen-based politics. They are rejecting the commodification of water and taking back control, becoming "keepers" of the fresh water systems in their localities. These reformers and fighters are the heroes and heroines of the story. Their courage and foresight shine in our hearts. If we follow their example, we may be able to save our vital supplies of fresh water before it is too late.

TREATY INITIATIVE

THE TREATY INITIATIVE TO SHARE AND PROTECT THE GLOBAL WATER COMMONS

We proclaim these truths to be universal and indivisible:

That the intrinsic value of the Earth's fresh water precedes its utility and commercial value, and therefore must be respected and safeguarded by all political, commercial, and social institutions,

That the Earth's fresh water belongs to the Earth and all species, and therefore must not be treated as a private commodity to be bought, sold, and traded for profit,

That the global fresh water supply is a shared legacy, a public trust, and a fundamental human right, and therefore, a collective responsibility,

And,

Whereas, the world's finite supply of available fresh water is being polluted, diverted, and depleted so quickly that millions of people and species are now deprived of water for life and,

Whereas governments around the world have failed to protect their precious fresh water legacies,

Therefore, the nations of the world declare the Earth's fresh water supply to be a global commons, to be protected and nurtured by all peoples, communities, and governments of all levels and further declare that fresh water will not be allowed to be privatized, commodified, traded, or exported for commercial purposes and must immediately be exempted from all existing and future international and bilateral trade and investment agreements.

The parties to this treaty — to include signatory nation-states and Indigenous peoples — further agree to administer the Earth's fresh water supply as a trust. The signatories acknowledge the sovereign right and responsibility of every nation and homeland to oversee the fresh water resources within their borders and determine how they are managed and shared. Governments all over the world must take immediate action to declare that the waters in their territories are a public good and enact strong regulatory structures to protect them. However, because the world's fresh water supply is a global commons, it cannot be sold by any institution, government, individual, or corporation for profit.

— *Written by Maude Barlow and Jeremy Rifkin and unanimously endorsed by the 800 delegates from 35 countries at the summit Water for People and Nature, Vancouver, July 8, 2001*

PART 1

THE CRISIS

RED ALERT

How the world is running out of fresh water

Water has been an important symbol in the legends and histories of many ancient cultures. Unlike people living in the urban, industrialized nations of the 21st century, most humans throughout history knew that their water resources could run out, and they developed a healthy respect for conserving whatever water they found. In biblical times, when Isaac returned to the land where his father Abraham had lived, the old wells he opened up were so important to life that they became a subject of dispute with other tribespeople. Later, Jacob's well was so highly prized and carefully protected that it was in use during the days of Jesus many centuries later.

Other societies, like the traditional Inuit and the early Mesopotamians, placed equal importance on the water that sustained the lives of their people. The Inuit depended largely on water-dwelling seals, fish, and walrus for their food, and their deity was a goddess of water, Nuliajuk. She ruled her realm with ferocious justice, and all of her power came from water. Nuliajuk gave the Inuit food from the sea and ice to build houses. When she withheld her gifts, no one could live. In the strikingly different world of the early Mesopotamians, water was treasured for different reasons. Before this group moved to the fertile valleys of northern Iraq, they lived in the dry plains of the south. They did manage to harness water for their farms, but it was very scarce. That is why their water-god, Enki, became one of the most important deities in their pantheon.

Thousands of miles away, in China, the dangers of drought became a theme of one myth, in which a Great Archer shot down nine out of ten suns, to prevent the earth from drying out. Chinese tradition also held that water and other elements of the earth exist in a balance that should not be disturbed. If there was a disruption in the normal cycles of Nature, Chinese governors were called upon to alleviate the problem. They were expected to help make up for the harm done to crops by reducing taxes or by distributing grain from the country's storehouses. Today, the normal cycles of Nature are being disrupted by climate change and the abuse of almost every water system on earth. However, unlike governments that followed the Chinese tradition described above, our governments are abdicating their responsibility to protect and conserve water, and they are handing its management over to the private sector.

Corporate control of the world's water resources and distribution systems is a threat to the well-being of humans around the world because water is fundamental to life. All living ecosystems are sustained by water and the hydrological cycle. Ancient peoples, and those living closer to the forces of Nature in today's world, knew that to destroy water was to destroy self. Only modern "advanced" cultures, driven by acquisition and convinced of their supremacy over Nature, have failed to revere water. The consequences are evident in every corner of the globe: parched deserts and cities, destroyed wetlands, contaminated waterways, and dying children and animals.

Nature is not entirely benign, and like the water-goddess of the Inuit, it will not tolerate this abuse forever. The signs are all present. If we do not soon change our relationship to water and the ecosystems that sustain it, all our wealth and knowledge will be meaningless. We are as dependent on fresh water for life as our ancient ancestors were. But many do not seem to be aware that this precious resource is disappearing. The clock is ticking, but they do not know it.

FINITE SUPPLIES

We'd like to believe there's an infinite supply of fresh water on the planet, and many of us have used water as if it would never run out. But the assumption is tragically false. Available fresh water amounts to less than one-half of one percent of all the water on earth. The rest is sea water, frozen in the polar ice, or water stored in the ground that is inaccessible to us. The hard news is this: humanity is depleting, diverting, and polluting the planet's fresh water resources so quickly and relentlessly that every species on earth — including our own — is in mortal danger. The earth's water supply is finite. Not only is there the same amount of water on the planet as there was at its creation; it is almost all the same water. Only a small amount may enter our atmosphere in the form of "snow comets" from the outer parts of the solar system. But even if the snow comet theory is correct, the speculated amount of water involved is so modest, it would do nothing to alleviate the shortage crisis.

The total amount of water on earth is approximately 1.4 billion cubic kilometers (about 330 million cubic miles). Canadian naturalist E.C. Pielou helps us visualize this statistic: if all the water on earth were solidified into a cube, each edge of the cube would be about 1,120 kilometers (about 695 miles) long, approximately twice the length of Lake Superior. The amount of *fresh* water on earth, however, is approximately 36 million cubic kilometers (about 8.6 million cubic miles), a mere 2.6 percent of the total. Of this, only 11 million cubic kilometers (about 2.6 million cubic miles), or 0.77 percent, counts as part of the water cycle in that it circulates comparatively quickly. However, fresh water is renewable only by rainfall. So in the end, humans can rely only on the 34,000 cubic kilometers (about 8,000 cubic miles) of rain that annually form the "runoff" that goes back to the oceans via rivers and groundwater. This is the only water considered "available" for human consumption because it can be harvested without depleting finite water sources.

Rain forms a crucial part of the hydrological cycle, the process through which water circulates from the atmosphere to the earth and back, from a height of 15 kilometers (about 9 miles) above the ground to a depth of 5 kilometers (3 miles) beneath it. Water that evaporates from the oceans and water systems of the continents goes into the atmosphere, creating a protective envelope around the planet. It turns into saturated water steams, which create clouds, and when those clouds cool, rain is formed. Raindrops fall on the earth's surface and soak into the ground, where they become groundwater. This underground water, in turn, comes back to the earth's surface in the form of sourcepoints for streams and rivers. Surface water and ocean water then evaporate into the atmosphere, starting the cycle anew.

Most of the earth's fresh water, however, is stored underground, just below the surface or deeper down. This is called groundwater, and it is 60 times greater in volume than the water that lies on the earth's surface. There are many types of groundwater, but the most important type for humans is "meteoric water" — moving groundwater that circulates as part of the water cycle, feeding above-ground rivers and lakes. Underground water reservoirs, which are known as aquifers, are relatively stable because they are secured in bodies of rock. Many of them are closed systems — that is, they are not fed by meteoric water at all. Wells and boreholes drilled into aquifers are fairly secure sources of water because they tap into these large reservoirs, but to be useful over time, an aquifer must be replenished with new water at approximately the same rate as the rate of extraction. However, around the world, people are extracting groundwater at rapid rates to supplement declining supplies of surface water.

MULTIPLE THREATS

All of the above-noted water sources are being taxed to their limit for multiple reasons. First, the world's population is exploding. Ten years from now, India will have an extra 250 million people and Pakistan's population will almost double, to 210 million. In five of the world's "hot spots" of water dispute — the Aral Sea region, the Ganges, the Jordan, the Nile, and the Tigris-Euphrates — the populations of the nations within each basin are projected to climb by between 45 and 75 percent by 2025. By that year, China will see a population increase greater than the entire population of the United States, and the world will house an additional 2.6 billion people — a 57 percent increase over today's level of 6.1 billion. To feed this many human beings, says the UN's Food and Agriculture Organization (FAO), agricultural production will have to increase by 50 percent. In such a scenario, demand for fresh water will obviously explode. As Allerd Stikker of the Amsterdam-based Ecological Management Foundation explains, "The issue today, put simply, is that while the only renewable source of freshwater is continental rainfall . . . [a finite amount of water], the world population keeps increasing by roughly 85 million per year. Therefore the availability of freshwater per head is decreasing rapidly."

Furthermore, increasing numbers of people are moving to cities, where dense populations place terrible strains on limited water supplies and make delivery of sanitation services next to impossible. For the first time in history, as many people now live in cities as in rural communities. There are 22 cities in the world with populations of over 10 million inhabitants. By 2030, says the UN, the world's cities will have grown 160 percent, and twice as many people will live in cities as in the countryside.

Second, as a result of many factors, per capita water consumption is exploding. Global consumption of water is doubling every 20 years, more than twice the rate of human population growth. Technology and sanitation systems, particularly those in the wealthy industrialized nations, have allowed people to use far more water than they need. The average Canadian household now consumes 500,000 liters of water every year (about 130,000 US gallons); each toilet — and many homes have more than one — uses 18 liters of water per flush (about five US gallons). And enormous amounts of water are lost through leakage in municipal infrastructure in countries all over the world. Yet even with the explosion in personal water use, households and municipalities account for only 10 percent of water use.

⌒

Industry claims the next big chunk of the world's fresh water supplies, at 20 to 25 percent, and its demands are dramatically increasing. Industrial use of water is predicted to double by 2025 if current growth trends persist. Massive industrialization is throwing off the balance between humans and Nature on many continents, especially in rural Latin America and Asia, where export-oriented agribusiness is claiming more and more of the water once used by small farmers for food self-sufficiency. Latin America and other Third World regions also host more than eight hundred free trade zones, where assembly lines produce goods for the global consumer elite, and these operations are another major drain on local water supplies.

Many of the world's growing industries are water intensive. It takes 400,000 liters (105,000 US gallons) of water to make one car. Computer manufacturers use massive quantities of de-ionized fresh water to produce their goods and are constantly searching for new sources. In the United States alone, the industry will soon be using over 1,500 billion liters (396 billion US gallons) of water and producing over 300 billion liters (79 billion US gallons) of wastewater each year. Originally thought to be a "clean" industry, high-tech has left a staggering pollution legacy in its short history. Silicon Valley has more Environmental Protection Agency (EPA) toxic Superfund sites than any other area in the U.S. and more than 150 groundwater contamination sites, many related to high-tech manufacturing. Close to 30 percent of the groundwater beneath and around Phoenix, Arizona, has been contaminated, well over half by the high-tech sector.

Irrigation for crop production claims the remaining 65 to 70 percent of all water used by humans. While some of this water use is for small farms, particularly in the Third World, increasing amounts are being used for industrial farming, which notoriously overuses and wastes water. These corporate farming practices are subsidized by the governments of industrialized countries and their taxpayers, and this creates a strong disincentive for farm operations to move to conservation practices such as drip irrigation. Much of the water usage that comes under this 65 percent heading should really be considered industrial, since modern factory farms have very little resemblance to community farms in any part of the world.

In addition to population growth and increasing per capita water consumption, massive pollution of the world's surface water systems has placed a great strain on remaining supplies of clean fresh water. Global deforestation, destruction of wetlands, the dumping of pesticides and fertilizers into waterways, and global warming are all taking a terrible toll on the earth's fragile water systems. (See Chapter 2.) Another source of pollution is the damming and diversion of water systems, which have been linked to unsafe concentrations of mercury and water-borne diseases. And many such projects are being constructed throughout the world. The number of large dams worldwide has climbed from just over five thousand in 1950 to forty thousand today, and the number of waterways altered for navigation has grown from fewer than nine thousand in 1900 to almost five hundred thousand. In the northern hemisphere, we have harnessed and tamed three-quarters of the flow from the world's major rivers to power our cities.

At the same time, overexploitation of the planet's major river systems is threatening another finite source of water. "The Nile in Egypt, the Ganges in South Asia, the Yellow River in China, and the Colorado River in America are among the major rivers that are so dammed, diverted, or overtapped that little or no fresh water reaches its final destination for significant stretches of time," warns Sandra Postel of the Global Water Policy Project in Amherst, Massachusetts.

In fact, the Colorado is so oversubscribed on its journey through seven U.S. states that there is virtually nothing left to go out to sea. The flows of the Rio Grande and upper Colorado rivers are in danger of being reduced by as much as 75 percent and 40 percent, respectively, over the next century, and in 2001, for the first time in recorded history, the Rio Grande ceased to flow into the Gulf of Mexico.

Water levels of the Great Lakes have also hit record lows in recent years. In 2001, the water was more than a meter below its seasonal average in the Port of Montreal, and Lakes Michigan and Huron were down by 57 centimeters (about 22 inches). Water flows in the St. Lawrence River are greatly affected by the water tables of the Great Lakes, and the environmental watchdog group Great Lakes United is warning that one day, the St. Lawrence may no longer reach the Atlantic Ocean.

DRYING PLANET

A powerful new study by hydrological engineer Michal Kravčík and his team of scientists at the Slovakian NGO People and Water shows in minute detail just how profoundly humanity's activities are affecting its sources of fresh water. Kravčík, who has a distinguished career with the Slovak Academy of Sciences, has studied the effect of urbanization, industrial agriculture, deforestation, paving, infrastructure building, and dam construction on water systems in Slovakia and its surrounding countries. He has come up with an alarming finding. Destroying water's natural habitat not only creates a supply crisis for people and animals, it also dramatically diminishes the *actual amount* of fresh water available on the planet.

Kravčík describes the hydrological cycle of a drop of water. It must first evaporate from a plant, earth surface, swamp, river, lake, or the sea, then fall back down to earth as precipitation. If the drop of water falls back onto a forest, lake, blade of grass, meadow, or field, it can cooperate with Nature and return to the hydrological cycle because it can be easily absorbed into soil or forest. But if it falls onto pavement and buildings in urban areas, it is not absorbed into the soil and instead it heads out to sea. This means that less water exists in the ground and rivers and less evaporates from land. Therefore a landlocked country will receive less rain because the water that should have stayed there (absorbed into the soil or rivers or lakes) has fled out to the ocean.

Kravčík explains that "the water cycle can be balanced if the volume of water flowing [from] the rivers [on] the continents into oceans equals the volume of water evaporated from the oceans, which comes back to the continents through frontal systems." However, sometimes there is a decrease in the amount of water moving down from the earth's surface and into the ground. This is called a drop in capillary action and it can be caused by overbuilt landscapes. When rain hits pavement and buildings instead of forest and soil, it cannot be absorbed and sent underground. Instead, it swells both rivers and oceans. As a result, precious fresh water is converted to salt water.

—

Kravčík's team also found that as the earth's surface is paved over — denuded of forests and meadows, and drained of natural springs and creeks — less precipitation is staying in river basins and continental watersheds, where it is needed, and more is heading out to sea, where it becomes salty. It is as if the rain is falling onto a huge, low-lying roof, or umbrella, of pavement and treeless areas: everything underneath stays dry, and the water runs to the perimeter. The water's forest and meadow "domicile" would have trapped falling rain and snow, but when it hits paved areas and denuded land, it slips off and heads out to the ocean. Kravčík believes the destruction of water-retentive landscapes is a serious violation. "Right of domicile of a drop of water," he says, "is one of the basic rights."

To quantify this theory precisely, the scientists studied Kravčík's own country, Slovakia, a small nation in central Europe that has undergone intensive urbanization in a very short time. The once rural countryside has been transformed into a "modern" state and its water systems have been radically altered to accommodate this passage. The scientists found clear evidence that all human interference in Slovakia's watersheds has caused faster outflow of rainfall water from the land to the oceans. They were actually able to quantify how water supplies decreased because of additional roofing, paving, car parks, and highways. Every year in Slovakia, about 250 million cubic meters (about 9 billion cubic feet) of fresh water disappear — one percent of all the water in Slovakia's watersheds. And since World War II, annual precipitation in Slovakia has decreased by 35 percent! Because of overbuilt landscapes, there are fewer places for water to congregate — such as wetlands and ponds — from which it could evaporate and then fall back as rain, right near the land that needs it.

Terrifyingly, the authors have been able to make some speculations about what this means globally. The world is urbanizing and therefore being paved over at about the same rate as Slovakia. This means that the continents are losing about 1,800 billion cubic meters (about 6,400 billion cubic feet) of fresh water a year, thus causing the oceans to rise by 5 millimeters (about a fifth of an inch) annually. If this trend continues, over the next hundred years, the land mass will lose about 180,000 billion cubic meters of fresh water, which is approximately equivalent to the volume of water of the whole hydrological cycle.

Kravčík's scientists have also issued a dire warning about the growing number of what they call "hot stains" on the earth — places where previously existing water has already disappeared. In the near future, the "drying out" of the earth will cause drought; massive global warming, with its attendant extremes in weather; less protection from the atmosphere; increased solar radiation; decreased biodiversity; the melting of polar ice caps; submersion of vast territories; massive continental desertification; and eventually, in Michal Kravčík's words, "global collapse."

In addition, a study published by the Scripps Institution of Oceanography at the University of California, San Diego, in November 2001 (partially funded by NASA), found that particles of human-produced pollution may be weakening the earth's hydrological cycle as well. Tiny aerosol particles made up of sulfates, nitrates, fly ash, and mineral dust produced by fossil fuel combustion are cutting down the amount of sunlight penetrating the ocean. The resulting reduction in heat means that less water evaporates back into the atmosphere, and less evaporation means less rain. As well, say the 150 blue-ribbon scientists who conducted the research, these aerosol particulates are suppressing rain over polluted regions as they trap water droplets in their web.

FRANTIC SEARCH

Not surprisingly, in the wake of the destruction of the world's surface fresh water supplies, communities, farmers, and industries are now aggressively seeking out the water supplies running free just under the earth's surface or held in deeper aquifer reservoirs. An estimated 1.5 billion people (about one-quarter of the world's total population) now depend on groundwater for their drinking water. Most areas of Asia, including the world's most populous countries — China and India — derive anywhere from 50 to 100 percent of their water supplies from groundwater. Some countries, such as Barbados, Denmark, and the Netherlands, are almost entirely dependent on this source. About one-third of the water used in France, Canada, and the United Kingdom is supplied by aquifers, while more than 50 percent of Americans are dependent on groundwater for their supplies. As a result of this explosive use of groundwater for day-to-day use around the world, massive groundwater overpumping and aquifer depletion are now serious problems in most of the world's most intensive agricultural areas and they are reaching critical levels in many of the world's large cities.

Aquifers vary enormously in size. As naturalist E.C. Pielou explains, for a layer of groundwater to function as an aquifer, it must be large enough to store a useful volume of water and permeable enough to be extracted at an acceptable rate. Aquifers are either *confined* (covered by a layer of rock or other sediment through which water cannot escape upward) or *unconfined* (saturated, so the trapped water goes right up to the level of the water table and a pipe can therefore be drilled down into the aquifer without going through rock or hard sediment). The most common method of searching for groundwater sources is to drill test wells or boreholes into the ground to search for new supplies. While wells have been used for centuries, extensive pump extraction of groundwater is a phenomenon of the late 20th century because of the availability of electricity and inexpensive equipment.

In many parts of the world, pump irrigation was originally seen as a godsend because it allowed crops to be grown year-round. It also made the controversial Green Revolution of Asia possible. This was a massive experiment, carried out in many Third World countries, including India, to make sure that every acre of workable land produced higher yields. To do this, monoculture replaced biodiversity, and great amounts of pesticides and fertilizers were used. While food yield did increase dramatically, the Revolution is now largely discredited because it destroyed biodiversity, increased chemical pollution, and relied on intensive irrigation. The Green Revolution also pitted farmer against farmer as they competed for water that they once shared and conserved according to traditional methods. It rendered traditional community ways of dealing with floods, drought, and water allocation obsolete. And the Green Revolution's dependency on intensive water use, as well as fertilizers and pesticides, sowed the seeds of its own failure.

Another problem with groundwater is that it can't be seen; farmers don't know an aquifer is gone until it suddenly dries up. In addition, massive groundwater extraction not only causes depletion of finite aquifer reserves, it dramatically reduces the water table of the whole surrounding area. When extraction exceeds recharge, the water also becomes progressively more expensive to pump and more contaminated with dissolved minerals. And crucially, because groundwater provides the principal source of water for streams, rivers, and lakes, these surface waters can also be depleted when aquifers are mined even if they do not dry up completely. River flows drop, ponds and marshes disappear, and salt water may invade emptied-out aquifers located in coastal areas. Water quality in the capital regions of Indonesia and the Philippines, for instance, has deteriorated sharply because of sea water intrusion. In some cases, aquifers emptied of water collapse in on themselves, especially if they are located under a large urban area. Thus, groundwater mining actually permanently reduces the earth's capacity to store water.

Pollution of underground water supplies has also become an issue as mining, manufacturing, and oil-extraction operations have expanded internationally. *World Resources*, a publication of the United Nations Environment Programme, reports that as Third World countries undergo rapid industrialization, heavy metals, acids, and persistent organic pollutants (POPs) are contaminating aquifers, often the only sources of local water.

~~

And in the Canadian province of Alberta alone, over 45 billion imperial gallons (about 204 billion liters) of water — much of it from aquifers — are pumped into oil wells every year to increase pressure in the reservoir and enhance production. This is enough fresh water to supply the seventy thousand residents of Red Deer for 20 years. Tragically, when the oil well is depleted, the water that remains behind is lost to people and Nature. It contains concentrated levels of minerals, as well as pollutants from the oil-drilling process.

Recently, oil companies and the Canadian government have invested heavily in the development of the Tar Sands — an oil sand reserve in northern Alberta the size of New Brunswick, which is estimated to contain about one-third of the world's remaining oil supplies, more than the reserves of Saudi Arabia. The process of separating the oil from tar sands requires huge volumes of water and is already diminishing stream and river flows in the area. Moreover, notes Canadian water expert Jamie

Linton, the process contaminates water to such a degree that it must be stored indefinitely in tailing ponds. Further, the deeper oil sands must be recovered by drilling horizontal wells and injecting steam far underground. This method of extraction requires nine barrels of water to produce one barrel of oil. Scientists predict severe water shortages in the region as a result.

Coal-bed methane production also involves withdrawing massive volumes of highly saline groundwater from coal-seam aquifers. An average well dewaters 16,000 US gallons (about 60,000 liters) of groundwater per day, discharging the saline water into rivers and streams and destroying aquatic life. In Montana alone, there are plans to develop between 14,000 and 40,000 coal-bed methane wells in the next decade. A mid-range estimate of 24,000 producing wells would pump 345 million US gallons of water per day (about 1.3 billion liters per day) from underground water reserves, lowering aquifer levels by 34 feet (about 10 meters) in ten years and causing massive saline pollution of the surrounding area.

Exponential increases in water use such as this have led the World Resources Institute to issue the following dire warning: "The world's thirst for water is likely to become one of the most pressing resource issues of the twenty-first century. . . . In some cases, water withdrawals are so high, relative to supply, that surface water supplies are literally shrinking and ground water reserves are being depleted faster than they can be replenished by precipitation." Put in economic terms, instead of living on fresh water *income*, we are irreversibly diminishing fresh water *capital*. At some time in the near future, we will be fresh water *bankrupt*.

PARCHED AMERICA

Although North Americans usually think of water shortages as a Third World problem, they are recently coming face to face with the crisis within their own borders. Twenty-one percent of irrigation in the United States is achieved by pumping groundwater at rates that exceed the water's ability to recharge, which means that aquifers like the Ogallala in the American Midwest are being rapidly depleted. As a result, farmers all over the region are reeling from a lethal combination of severe drought and dried-up wells.

And the cost of losing American farmland because of the depletion of aquifers is over US$400 billion every year.

The Ogallala aquifer is probably the world's most famous underground body of water. It is the largest single water-bearing unit in North America, covering more than half a million square kilometers (about 190,000 square miles) of the American High Plains regions. It stretches from the Texas panhandle to South Dakota and is believed to contain about 4 trillion tons of water — 20 percent more water than Lake Huron in the Great Lakes. Although it is made up of fossil water — water locked deep underground for thousands of years with few sources of replenishment — it is being mined mercilessly by over 200,000 wells irrigating 3.3 million hectares (about 8.2 million acres) of farmland — one-fifth of all the irrigated land in the United States. At a withdrawal rate of 50 million liters (about 13 million US gallons) a minute, water in the Ogallala aquifer is being depleted 14 times faster than Nature can restore it. Since 1991, each year the water table in the aquifer has dropped by at least a meter (about three feet) — a huge amount when multiplied by the aquifer's area. By some estimates, more than half of its water is already gone.

The destruction of the Ogallala aquifer is probably America's most notorious headlong rush into water scarcity, but many other regions in the country are depriving themselves of water security as well. California, for instance, is in big trouble. Its aquifers are drying up, the Colorado River is strained to the limit, and the water table under California's San Joaquin Valley has dropped nearly 10 meters (about 33 feet) in some spots within the last 50 years. Overuse of underground water supplies in the Central Valley has also resulted in a loss of over 40 percent of the combined storage capacity of all human-made surface reservoirs in the state. California's Department of Water Resources predicts that, by 2020, if more supplies are not found, the state will face a shortfall of fresh water nearly as great as the amount that all of its towns and cities together are consuming today.

Population continues to explode in the deserts of the American Southwest, which are largely barren of water. More than eight hundred thousand people live in greater Tucson alone and four million in all of Arizona, a tenfold increase in 70 years. Until recently, Tucson relied solely on aquifers for its water. But as overextraction increased, the depth of the wells grew from 150 meters to 450 meters (about 490 to 1,500 feet), and the city started importing supplies from the Colorado River and buying up local farms for their water, taking farmland out of production. Municipal development in Phoenix is occurring at a rate of an acre every hour, so it's small wonder that water tables have dropped more than 120 meters (about 390 feet) east of Phoenix. Projections for Albuquerque, New Mexico, show that if groundwater withdrawals continue at current levels, water tables will drop an additional 20 meters (about 66 feet) by 2020 and major cities in the region will go dry within 10 to 20 years.

Even in the suburbs around rainy Seattle, demand for water is outstripping supply, with predicted shortages in 20 years. In the much drier El Paso, Texas, all current sources of water are expected to be gone by 2030, and in northeast Kansas, the water shortage is so severe that state officials are discussing plans to build a pipeline to the already overtapped Missouri River. Similarly overtaxed is the huge sandstone aquifer lying under the Illinois-Wisconsin border, the source of water for millions of people, including the populations of Chicago and Milwaukee. A hundred years of pumping have mined this source relentlessly, and for decades, scientists have been monitoring the decline of the aquifer's water table, warning that, unless groundwater withdrawals are reduced, it will definitely run out of water in the foreseeable future.

Farther east, in Kentucky, more than half the state's 120 counties ran short of water during the summer of 2001. And on the Atlantic seaboard, Long Island takes water from a closed-basin aquifer rapidly being depleted and poisoned by industrial runoff. Meanwhile, the Ipswich River in Massachusetts is running thin, and Eastern cities, such as Philadelphia and Washington, whose water is notoriously bad, are searching farther afield for secure long-term water sources.

Like the Ogallala aquifer, the Florida aquifer system in the Southeast is also being mined far faster than it can naturally be replenished. Though it is about 200,000 square kilometers (about 76,000 square miles) in area and extends under several states besides Florida, its water levels are dropping dangerously low as its water is being extracted at a rate of 6.6 million liters (about 1.7 million US gallons) per minute. The water table has plummeted so far in Florida that sea water has invaded its aquifers. Incredibly, Florida Governor Jeb Bush is championing a proposal to collect surface water and inject it, untreated and contaminated by all sorts of impurities, back into the depleted groundwater sources.

DESPERATE MEXICO

South of the American border, the problem gets worse. Mexico City was once an oasis — an Aztec center called Tenochtitlan, which was literally an island city ringed by lakes and connected to the mainland by three causeways. Crisscrossed by abundant canals, aqueducts, dikes, and bridges, it was also a haven of floating gardens and baths. When the Spanish invaded in 1521, they tore down all the great Aztec buildings, destroyed the dikes, and using an endless supply of local slave labor, filled in and drained the lakes. Orders had been given that Mexico City, the capital of New Spain, should resemble a great Spanish city, not Venice. The protective forests surrounding the area were destroyed as well.

For five centuries, Mexico City's population remained static. The total in 1845 was only 240,000. Then it suddenly started to grow. It passed the million mark in 1930 and stands today at a breathtaking 22 million. Poor urban planning resulted in endless expansions of concrete, which covered the remaining drainage and free-flowing water, and an estimated 40 percent of its piped water leaks from a crumbling infrastructure built a hundred years ago. When the rain falls, it has nowhere to go but into the enormous subterranean system where it mixes with raw sewage and is pumped out of the city to irrigate the adjacent farmlands.

The pressure on the region's groundwater sources is understandably relentless. Mexico now depends on aquifers for 70 percent of its water and is extracting from aquifers 50 to 80 percent faster than the rate of regeneration. About a third of the city's water has to be pumped up to an area 2,300 meters (about 7,500 feet) above sea level, some from as far as 300 kilometers (about 180 miles) away. Mexico City is literally running out of water; experts are saying the city could go completely dry in the next ten years.

For decades, the city has also been sinking as underground water pockets have been replaced by air. The process, familiar to those living next to coal or oil extraction, is called subsidence. Mexico City was the first to experience the phenomenon as a result of water removal, because it sits on a porous, sponge-like subsoil. The more water people drink in Mexico City, the more they sink down into their foundations. Old sewer and water pipes are being crushed and architectural treasures are cracking and teetering. The city has sunk steadily into the mud for decades and is now subsiding at a rate of about 50 centimeters (about 20 inches) annually.

The crisis is not confined to the Mexican Valley. Years of drought in the northwestern state of Sonora have left the region dry as a bone, and Sonora's Batuc Reservoir, created when the Moctezuma River was dammed 35 years ago, is empty, eerily exposing a chapel and cemetery that were submerged at the time. North of Sonora, all along the Mexico-U.S. border, the export-processing zones known as the *maquiladora* employ millions of young Mexicans at slave wages, in unsafe, toxic conditions. Here, fresh water is so scarce that it is delivered weekly in many communities by truck or cart. Ciudad Juárez, growing at a rate of fifty thousand people a year, is running out of water, and the underground aquifer the city relies on has declined at about five feet (about one and a half meters) per year. At this rate, there will be no usable water left in 20 years.

MID-EAST CRISIS

Almost every country in the Middle East is facing a water crisis of historic proportions. In the Arabian Peninsula, groundwater use is nearly three times greater than recharge, and at the current rate of extraction, Saudi Arabia, which depends on aquifers for 75 percent of its water, is running toward total depletion in the next 50 years. In an attempt to become food self-sufficient, the country subsidized farmers to extract water, but this benefit came at a terrible cost. For every ton of grain produced, three thousand tons of water were used — three times the norm. Aquifer depletion curtailed this project, but not before the country's fresh water supplies were devastated. And in Iran, people are suffering through the worst water shortage in decades. The official IRNA news agency reports that Iran's farms are short 1.2 billion cubic meters (about 39 billion cubic feet) of water. Severe droughts are intensifying the crisis.

In Israel, extraction has exceeded replacement by 2.5 billion cubic meters (about 88 billion cubic feet) in 25 years, and 13 percent of the country's coastal aquifer is contaminated by sea water and fertilizer runoff. According to its own officials, Israel will have a water deficit of 360 million cubic meters (about 12 billion cubic feet) by 2010, but already in July 2001, the Israeli government announced that the country faced "the deepest and most severe" water crisis ever as three years of drought have led the government to consider a nationwide ban on watering lawns. Water Commissioner Shimon Tal warned that the country would have to live "hand to mouth" until planned desalination plants begin operating and taking water out of the sea.

Israel gets about half of its water from Lake Kinneret (the Sea of Galilee), which is fed by the Jordan River, but the lake has experienced perilously low water levels in recent years and is beginning to be infiltrated by saline water. Most of the rest of Israel's water comes from two aquifers — the Mountain and Eastern aquifers — which, between them, supply most of the water for residents and farmers in the controversial settlement areas of the West Bank and the Huleh Valley. In the latter region, which was in Syrian territory before the 1948 war, massive farming operations based on aquifer mining have devastated water sources. In his book *Water*, Marq de Villiers describes how this came about. Wetlands were drained; groundwater tables began to drop; and streams and springs dried up. To deal with new salinity in the water, resulting from salt left behind when water systems were depleted, farmers switched to salt-resistant crops, but to no avail. The aquifers dried up and the earth subsided into the cavities left behind, just as it is doing in Mexico City. Some of the air pockets were so huge that whole houses were swallowed up and disappeared.

Palestine and Jordan are experiencing similar devastation. Palestine's Gaza Strip has one of the highest population growth rates in the world and relies almost exclusively on groundwater. However, salt water intrusion from the Mediterranean has been detected as far as a mile inland, and some experts predict that the country's groundwater will become entirely salinized. In Jordan, the sole source of surface water is the Jordan River, and when Israel began diverting it for irrigation projects in the south of Israel, its water levels dropped. They are now only one-eighth of what they were 50 years ago, and this has forced Jordan to mine its limited aquifer systems to overcapacity. Groundwater in the country is now being used up 20 percent more quickly than the recharge rate. A tragic by-product of the diversion of the Jordan River is its impact on the Dead Sea. As Friends of the Earth, Middle East, explain, the surface of this body of water has dropped more than 25 meters (82 feet) in the last three decades and the drop is accelerating. The Dead Sea is dying, says the group. Its entire southern basin is dried up and has been transformed into an industrial site, and life-threatening sinkholes have appeared along its shoreline.

~

Elsewhere in Jordan, one underground water system with great symbolic meaning to Jordanians has been devastated. The Oasis of Azraq, deep in the Jordanian desert, has for centuries been a resting place for animals, migrating birds, and humans — a wondrous water-filled sanctuary fed at one time by more than ten subterranean springs. This oasis was so important to Jordan that it was named as an international wetland heritage site in 1977. However, desperate for water, the Jordanians started pumping from the Azraq 20 years ago, sending about 900 cubic meters (about 32,000 cubic feet) an hour to Amman, the capital. Within a few years, many wells had been built and were pumping almost three times that amount of fresh water, double what the basin can sustain. As Alanna Mitchell of the *Globe and Mail* reported, by 1993, the oasis was a dusty garbage dump, the land an open sore of deep fissures from which a searing heat arises.

Unfortunately, lessons from these terrible stories haven't changed humanity's behavior. Libya, which has used up all its conventional water sources and exposed its coastal aquifers to excessive mining, decided a decade ago to mine the sub-Saharan aquifer that lies under parts of Chad, Egypt, Libya, and Sudan. Known as the Nubian Aquifer, it is one of the world's most extensive. At the same time, at an estimated cost of over US$32 billion, Libya has hired a huge South Korean conglomerate to construct a 1,860-kilometer (1,000-mile) pipeline to take fresh water from the aquifers of the Kufra Basin in the Sahara Desert and use it to support the farms and cities in the northern part of the country. Much of the project has been completed, and nearly one thousand wells now take water from beneath the desert.

Already, over one billion cubic meters (about 35 billion cubic feet) of water a year are being mined; when fully operational, the volume of water pumped from the aquifer will be 40 billion cubic meters (about 1,400 billion cubic feet) a year — equal to the flow of any great river. Libya's head of state, Moammar Gadhafi, variously calls this project the "Great Man-Made River" and the "Eighth Wonder of the World." At this rate, the aquifer could be empty in 40 to 50 years, affecting not only Libya, but all the countries around it.

CHINA'S "MIRACLE"

Perhaps the most disturbing reports of water crisis come from the country with the largest population on the planet. China has almost one-quarter of the world's population but only 6 percent of its fresh water. All over the country, wells are mysteriously emptying out, water tables are dropping, and rivers, streams, and lakes are drying up. As large industrial wells probe the ground ever deeper to tap the remaining water, millions of Chinese farmers have found their wells emptied. The western half of China is made up mostly of deserts and mountains, and the vast bulk of the country's 1.2 billion citizens live on several great rivers whose systems cannot sustain demands. For instance, in 1972, the Yellow River failed to reach the sea for the first time in history. That year it failed on 15 days; every year since, it has run dry for a longer period of time. In 1997, it failed to reach the sea for 226 days. The story is similar with all of China's rivers.

Water tables on the North China Plain — China's breadbasket — are dropping 1.5 meters (about 5 feet) a year, and northern China now has eight regions of aquifer overdraft. Four hundred of the country's six hundred northern cities are already facing severe water shortages, as is over half of China's population. And though water previously used by millions of farmers has been diverted to Beijing by deliberate government policy, the water table beneath the capital city has dropped 37 meters (about 120 feet) over the last four decades. The projected water crisis in Beijing is so severe that experts are now wondering whether the seat of power in China will have to be moved.

These shortages come at a time when conservative estimates predict that annual industrial water use in China could grow from 52 billion tons to 269 billion tons in the next two decades, and when rising incomes are allowing millions of Chinese to install indoor plumbing with showers and flush toilets. The Worldwatch Institute predicts China will be the first country in the world that will have to literally restructure its economy to respond to water scarcity.

The Worldwatch Institute also warns that an unexpectedly abrupt decline in the supply of water for China's farmers could threaten world food security. In the near future, China will experience severe grain shortages because limited water resources are currently being shifted from agriculture to heavier industrial and urban users. Planners in China estimate that a given amount of water used in industry generates more than 60 times the cash value of the same water used in agriculture, and political leaders have responded by diverting more and more of China's rural water sources to its burgeoning industrial base. But when China faces shortfalls in its own grain production, the resulting demand for imported grain could at times exceed the world's available exportable supplies. China might be able to survive such shortages for a time because its booming economy and huge trade surpluses will give it the cash needed to buy grain. However, this rising demand will force the prices of imported grain to go up. And this, in turn, will create social and political upheaval in many major Third World cities and threaten global food security.

SPREADING DISASTER

The story has been repeated in many other countries and regions. Most African countries start with a limited water supply, which is then stretched even further by drought, population growth, and pollution. Africa, already home to the most sprawling desert in the world — the African Sahara — continues to suffer from desertification. Vast nonrecharging aquifers underlie this desert, but these are connected to the water mining project that Libya's Moammar Gadhafi has set up for his country. Current depletion of these aquifers is estimated at 10 billion cubic meters (about 352 billion cubic feet) a year, and these rates will only increase as each stage of the project is completed.

According to Marq de Villiers, as many as 22 African countries fail to provide safe water for at least half their population: Guinea-Bissau, Guinea, Sierra Leone, Sao Tome and Principe, Mali, Niger, Nigeria, Cameroon, Congo, the Democratic Republic of Congo, Angola, Lesotho, Swaziland, Burundi, Mozambique, Madagascar, Uganda, Kenya, Ethiopia, Somalia, Djibouti, and Eritrea. However, it is India that has the highest volume of annual groundwater overdraft of any nation in the world. In most parts of the country, water mining is taking place at twice the rate of natural recharge, causing aquifer water tables to drop by 3 to 10 feet (about 1 to 3 meters) per year. Especially hard hit are the Punjab and Haryana states, India's breadbasket, and the northwestern state of Gujarat, where 90 percent of the wells have experienced a serious decline in water level. In the state of Tamil Nadu, groundwater tables have fallen as much as 99 feet (about 30 meters) in 30 years, and many aquifers have run dry. In the state of Rajasthan, the water system of the city of Jodhpur literally exploded when the water table beneath the city was drained dry. And in the Punjab and in the country of Bangladesh, the drop in the water table is even greater than China's, even though those places experience flooding every year. According to the International Water Management Institute, a quarter of India's grain harvest could be lost in the near future because of aquifer depletion.

RED ALERT

According to the United Nations, 31 countries in the world are currently facing water stress and scarcity. Over one billion people have no access to clean drinking water and almost three billion have no access to sanitation services. By the year 2025, the world will contain 2.6 billion more people than it holds today, but as many as two-thirds of those people will be living in conditions of serious water shortage, and one-third will be living with absolute water scarcity. Demand for water will exceed availability by 56 percent.

Many of us who have lived most of our lives in the industrialized countries of the North may find it difficult to imagine running out of water. We have lived with steady supplies most of our lives and have used it lavishly. But at current rates of use, we will run short. At a time when we are on a rising curve in water use because of increasing industrialization, intensified farming, and population growth, water resources are being depleted at an accelerated rate. Aquifer overdrafts, massive urbanization, and unchecked pollution are withdrawing supplies from the world's water account, just when we need to be saving more. And as we shall see in the next chapter, wetland loss, toxic runoff, and other forms of environmental damage are also threatening the world's precious remaining supplies of water. There is simply no way to overstate the fresh water crisis on the planet today. The alarm is sounding. Will we hear it in time?

ENDANGERED PLANET

*How the global water crisis is
endangering the planet and
other species*

Canadian environmentalist David Suzuki explains the concept of "exponential environmental destruction" to audiences all over the world.
Environmental problems, he says, don't grow in linear fashion, one step
at a time. None of us can see all of them, or even most of them. An ecosystem may be attacked in a thousand ways, from a thousand sources.
But because the assault is not a strict progression you can monitor, the
ecosystem might look fine one day and be dead the next. It's not two plus
two, four plus four, or eight plus eight, but two times two, four times four,
sixteen times sixteen, and so on.

Suzuki uses a riddle to illustrate his point. He asks his audience to
imagine a lake with a water lily on it. He explains that water lilies are great
as long as they don't get too prolific. If they stay under control, they and
the lake can coexist very well. But if the water lilies cover the lake entirely,
oxygen will be cut off, and the lake will die. Now, says Suzuki, we have a
lake and a water lily. In 60 days this lily will have grown exponentially, and
on day 60, lilies will totally cover the lake and it will die. What, he asks,

does the lake look like on day 59? The answer is that on day 59, the lake is only half covered with water lilies and looks fine.

If environmental destruction was taking place one step at a time, there would be as much time ahead to fix the problem as it took to create it. One plus one. You could count the problems each day and assess the danger. But when the devastation is exponential, the cumulative effect of all the sourcepoints hits all at once, and not always with warning. Applying this analysis, David Suzuki might say that, in terms of water security, the planet is on day 59.

———

Fresh water systems are both disproportionately rich and disproportionately imperiled. Although occupying a small area compared to land and oceans, they are home to a relatively high proportion of species, with more per unit area than other environments — 10 percent more than land and 150 percent more than oceans. Twelve percent of all animal species, including 41 percent of all recognized fish species, live in the less than one percent of the earth's surface that is fresh water. Yet over the last several decades, at least 35 percent of all fresh water fish species have become extinct, threatened, or endangered, and entire fresh water fauna systems have disappeared. Fresh water animals in North America are five times more likely to be at risk of extinction than animals that live predominantly on land.

Even more disturbing is the *rate* at which species are being lost. The highly regarded journal *Science* reports that recent extinction rates are one hundred to one thousand times higher than before humans existed, and if nonthreatened species become extinct by the end of this century, the loss of those species will speed up extinction rates overall — to one thousand to ten thousand times prehuman levels. According to Smithsonian Institution biologist Jonathan Coddington, whom Janet Abramovitz quotes in her article "Sustainable Freshwater Ecosystems," there will be a "biodiversity deficit," whereby species and ecosystems will be destroyed at rates faster than Nature can create new ones.

This catastrophe has not just "happened." It is partly the accumulated result of a set of human-driven assaults on the planet's fresh water systems — assaults that continue every day.

TOXIC RUNOFF: SEWAGE AND CHEMICALS

The single biggest threat to fresh water species is pollution from thousands and thousands of factories, industrial farms, and cities that pour or leak pesticides, fertilizers and herbicides (including nitrates and phosphates), bacteria, medical waste, chemicals, and radioactive wastes into our water. There, they add excess organic matter and nutrients such as nitrogen and phosphorus, which create algae, which in turn, rob the water of its oxygen. They also contribute disease-carrying pathogens such as cryptosporidium and sediment that smothers habitat. The rate at which oxygen is used up by algae is called the biomedical oxygen demand (BOD), and this is used to measure pollution. The entire process is called *hyper-trophication*, or "galloping eutrophication."

Some pollutants come to the water through the air. They enter the atmosphere from industrial smokestacks and from vehicle exhaust. Acid rain is created when some of these industrial gases, such as sulfur and nitrogen oxides, dissolve in falling rain. That rain then falls on surface water, which in turn becomes acidified, killing the lake and everything in it. In some Canadian lakes, acid rain has caused a 40 percent decline in fish species. However, as Canadian naturalist E.C. Pielou explains, acid rain is not the only cause of acid surface water. Acidic drainage from coal and mineral-ore mines produces sulfides, used for producing sulfuric acid commercially. Unfortunately, these sulfides also combine with oxygen and water naturally and leak sulfuric acids from the ground into lakes and streams.

Pollutants enter groundwater in many ways. Leaky gasoline tanks and municipal sewage lagoons, municipal landfills, feedlot effluent, mine tailings, septic tank ruptures, oil spills, pesticide runoff, and even road salt are all sources of groundwater pollutants. They form *leachate*, which is carried into groundwater when it rains. Unconfined aquifers are most affected as the pollution can enter them more easily and spread more quickly through the water table. Some pollution, such as gasoline, is lighter than water and will sit on the top of an aquifer or run along the top of an underground river. From there, it will leak benzine and other chemicals in a plume of pollution through the aquifer. Meanwhile, heavy polluting liquids might sink through the same aquifer and settle at the bottom.

Some of these heavy pollutants are extremely powerful. For example, a standard 200-liter (53-US-gallon) drum of the oily industrial solvent trichloroethylene would need to be diluted with 60 billion liters (about 16 billion US gallons) of water to make it harmless. Another lethal heavy pollutant is methyl tertiary butyl (MTBE), a methanol-based gasoline additive. Although it is known that a few drops of MTBE can contaminate a mid-sized aquifer, this chemical has been found leaking into over ten thousand wells throughout the state of California.

The National Geographic reports that one billion pounds (about half a billion kilograms) of industrial weed and bug killers are used throughout the United States every year, and most of it runs off into the country's water systems. Because of pollutants like this, nearly 40 percent of U.S. rivers and streams are too dangerous for fishing, swimming, or drinking, and fish and other water-dwelling wildlife have become living toxic-waste carriers. Thirty-seven percent of fresh water fish are at risk of extinction, 64 percent of crayfish and 40 percent of amphibians are imperiled, and 67 percent of fresh water mussels are extinct or vulnerable to extinction. "We have crashing ecosystems in every river basin in the West," says the Sierra Club's Colorado River Task Force.

Along the Mexico-U.S. border, people work for incredibly low wages, producing goods for global markets in what are known as *maquiladora* free trade zones. These areas are thick with industrial and human waste, but only about one-third of the wastewater and sewage going into nearby rivers and streams is treated. As a result, one environmental group calls the Mexico-U.S. border a "3,400 kilometer Love Canal." In his book *The Corporate Planet*, Josh Karliner describes the destruction of the fresh water systems of the region. Because the New River, stretching from Baja California in Mexico to the Imperial Valley in the U.S., is laden with more than a hundred toxic chemicals, U.S. health officials warn people not to even go near the deadly body of water. One government study found that 75 percent of all *maquiladora* factories were dumping toxic waste directly into rivers and streams, and yet families actually live on rivers in the area that are swollen with this poisonous industrial waste, as well as garbage, industrial runoff, and even the carcasses of animals killed by drinking the water.

TOXIC WORLD WATER SYSTEMS

Most of the world's waterways are now struggling with the full range of modern industrial toxic pollution problems, and there seems to be no end in sight. According to the United Nations Industrial Development Organization (UNIDO), industrial activity is likely to consume twice as much water by the year 2025, and industrial pollution is likely to increase fourfold. Untreated sewage is also killing waterways all around the world. Ninety percent of wastewater produced in the Third World is still discharged, untreated, into local rivers and streams. Africa's Lake Victoria is imperiled by the dumping of millions of liters of raw sewage and industrial waste from the cities of surrounding Kenya, Tanzania, and Uganda, and the fish stock of the Senegal and Niger Rivers is nearly depleted. In China, 80 percent of the major rivers are so degraded that they no longer support fish. The Yangtze River is contaminated with 40 million tons of industrial waste and raw sewage every day, and the water in the Yellow River is so polluted that it cannot be used even for irrigation. China's rivers are laced with intensive concentrations of human waste.

The Ganges and the Brahmaputra of India are similarly filled with bacteria and high fecal count, and nearly 200 million liters (about 53 million US gallons) of untreated sewage pour into the Yamuna River from Delhi's sewage system every day. The river is now considered to be irreparably damaged, as is the Damodar, which is filled with toxic sludge from the industries that edge its banks. India is home to the most polluted water in Asia, outside of China. The coasts of Bombay, Madras, and Calcutta are putrid. The sacred Ganges, where millions come to purify themselves, is an open sewer.

In Japan, water pollution comes from heavily chlorinated solvents from industry. In Jakarta, Bangkok, and Manila, indiscriminate dumping of liquid effluents and solid wastes has led to outbreaks of cholera, typhoid, and other water-borne diseases. The Mekong River, which begins in China and drains through Myanmar (formerly Burma), Laos, Cambodia, and portions of Thailand and Vietnam, is choking with industrial and human waste.

An alarming number of rivers and lakes in eastern Europe are ecologically dead or dangerously polluted. Three-quarters of Poland's rivers are so contaminated by chemicals, sewage, and agricultural runoff that their water is unfit even for industrial use. The same is true of the rivers in the Czech Republic and Slovakia. Sofia, the capital of Bulgaria, ran so low on water in 1995 that its citizens could turn on their taps only every second or third day. Nearly half the water and sewage treatment systems in Moscow are ineffective or malfunctioning, and according to the Russian Security Council, 75 percent of the republic's lake and river water is unsafe to drink.

Elsewhere in Europe, well-known rivers are failing. England's 33 major waterways are losing volume as a result of overuse of water, and some are now less than a third of their average depth. A hundred years ago, 150,000 salmon were caught annually in the Rhine River in the Netherlands and Germany alone, but by 1958, the salmon had disappeared. Development has also cut the Rhine off from 90 percent of its original flood plains, and the river's banks are lined with 20 percent of the world's chemical factories. It runs through the most heavily populated and industrialized parts of Europe, and much effluent is still discharged directly into it. To the southeast, the "blue Danube" carries a load of phosphates and nitrates that have increased sixfold and fourfold, respectively, over the last 25 years, causing great harm to the region's tourism and fisheries. All these rivers eventually transport their effluents to the sea, and many end up in the Mediterranean. There they create a breeding ground for invasive species and deadly algae. In recent years, a deadly algae called *Caulerpa taxifolia* has spread throughout the Mediterranean Sea at about four hectares (about 10 acres) a day and is threatening marine life along the entire coast.

Even rainwater is not pure anymore in Europe. Researchers from the Swiss Federal Institute of Environmental Science and Technology reported recently that rainfall on the continent is so full of toxic pesticides that much of it is too dangerous to drink. As in North America, much of the water bottled for commercial sale is taken from sources contaminated with industrial pollution and human and animal waste.

In Canada, a wealthy country with abundant supplies of water, over one trillion liters of untreated sewage are dumped into waterways every year. This volume would cover the entire 7,800-kilometer (4,800-mile) length of the Trans-Canada Highway to a depth of 20 meters (about 66 feet) — six stories high. And in industrialized nations like Canada, sewage is no longer just human waste. In a 2001 study, the Sierra Legal Defence Fund explains the problem. "The information we uncovered," the Fund wrote in a published study called *The National Sewage Report Card (Number Two)*, "is particularly disturbing when you consider what sewage really is — a foul mix of water, human excrement, grease, motor oil, paint thinner, antifreeze and many kinds of toxic industrial and household waste." And even treated waste can be lethal. While treatment will remove fecal coliform bacteria, the best-known variety of which is the deadly *E. coli*, it does not remove the toxic chemicals contained in wastewater. A July 2001 study by Quebec's Environment Ministry found that water flushed into the province's lakes and rivers is still "acutely toxic," even after highly sophis-ticated treatment. Pesticides, industrial wastes, arsenic, and metals all showed up in the "treated" water flowing into the St. Lawrence River. The Quebec study observed that "in all, more than 85 percent of the sewage samples from all sources contained the following: Ammonia, phosphorus, aluminum, arsenic, barium, mercury, PCBs, chlorinated dioxins, and furans, surfactants (cleaning chemicals), polyaromatic hydrocarbons (PAHs), and other organic and inorganic wastes." More than one in five samples of post-treated water flowing into the St. Lawrence was contaminated enough to kill the rainbow trout forced to swim in it. In a similar study in Ontario, more than half the trout were killed after swimming in the water samples. And the 2001 Sierra study found that "typical municipal water contains some 200 synthetic chemicals," including PCBs, which make water danger-ous to drink. As the researchers noted, "Just one drop of oil can render 25 liters [6.6 US gallons] of water unsafe for drinking. One gram [0.04 ounces] of polychlorinated biphenyls (PCBs), a substance used in everything from cosmetics to pesticides, is enough to make one billion liters [about a quar-ter of a billion US gallons] of water unfit for freshwater life."

In spite of the fact that synthetic chemicals are rendering waterways unsafe for human consumption, the volumes entering the environment are not being reduced. In fact, the use of chemicals has exploded in the last several decades. And every year, nearly US$2 trillion worth of chemicals are manufactured worldwide, and most of them find their way into our water. In the free trade zones of Mexico, for instance, the production of toxic chemicals has tripled since NAFTA was signed in 1994. Every year, the 1,200 plants of Baja California on Mexico's Pacific coast produce 36,000 tons of toxic residue. San Diego County produces even more — 160,000 tons in 2000. It's not surprising that all North Americans are carrying at least five hundred chemicals in their bodies that were unknown before World War I.

Another chemical hazard to watersheds is the effluent that runs into lakes and rivers from pulp and paper mills. The pulp and paper industry uses huge volumes of water and dumps oxygen-depriving effluent into waterways that then become choked with algae. Most mills also use chemicals to break wood down into paper. The most harmful ones produced in the chlorine-bleaching process are dioxins and furans, some of the deadliest known toxins in the world, and they contaminate surface water and groundwater alike. And pulp and paper processing is not a minor presence in Canada. It is estimated that the pulp and paper industry is responsible for fully half of all the waste dumped into Canada's waters.

Unlike the forestry industry, farming has traditionally been considered a relatively benign influence on the environment. But large-scale farms now mass-produce animals, confining them in crowded feedlots and factory-style barns. These operations create a staggering amount of manure — more than 130 times the amount of human waste produced in the United States. Texas alone creates an estimated 280 billion pounds (127 billion kilograms) of manure annually — 40 pounds (18 kilograms) per Texan. In factory-farm operations throughout North America, millions of gallons of liquefied animal feces are stored in open lagoons that emit over 400 different volatile, dangerous compounds into the atmosphere. These "sewerless cities" generate so much surplus manure that it cannot be stored or disposed of safely. Some large hog farms produce volumes of untreated hog manure equivalent to the human waste of a city of 360,000 people.

This waste, laced with antibiotics, is finding its way into groundwater and surface water in huge quantities. In some cases, it leaks; in others, it spills. In 1998, reports David Brubaker of the Center for a Livable Future at Johns Hopkins University, a 100,000-gallon (about 380,000-liter) spill in Minnesota killed 700,000 fish, and in Indiana in 1997 alone, there were over two thousand such spills. In the summer of 2000, more than one hundred of these lagoons were destroyed by Hurricane Mitch in North

Carolina, turning much of the state into a toxic mess. That same summer, hog waste spilled from the Great Lakes Swine Farm in Palmyra, Ontario, and leaked into Lake Ontario. And in California, factory-farm runoff is leaking into the already overtaxed Ogallala aquifer. Toxic sludge also finds its way into water when it is sprayed over farm fields, an increasingly common practice. (In Canada, this method is even being used to spread human waste!)

If farms like this were to stop producing such incredible amounts of concentrated waste, they would still have to deal with the massive quantities of nitrogen that they inject into the environment in the mass production of food. Intensive farming uses such high concentrations of nitrogen fertilizers that the practice has destabilized Nature's nitrogen balance and fouled water sources. In its natural state, nitrogen is an innocuous molecule that makes up 79 percent of the air we breathe. As the Minneapolis-based Institute for Agriculture and Trade Policy explains, before human dominance of the world's ecosystems, the primary sources of nitrogen were biologically natural and the earth adapted to recycling the chemical efficiently. Little excess nitrogen existed. But the massive use of nitrogen fertilizers and other manufactured sources of nitrogen has pushed twice as much of the chemical into the environment as was found there before these inputs were used.

The doubling of nitrogen in water and soil cycles has had a profound impact on the world's ecosystems. Excess nitrogen in water lowers the level of oxygen, which in turn affects the metabolism and growth of oxygen-dependent species. This leads to a condition called *hypoxia*. In one horrific example of the effect of synthetic fertilizers on water systems, many of the nitrates spread on farm fields all over the American Midwest do not stay there, but leach instead into the Mississippi River through streams and tributary rivers. The sum total of all the nitrogen runoff then moves down the river and heads out to the Gulf of Mexico, where it has created a "dead zone" of 18,000 square kilometers (about 6,900 square miles) — about the size of New Jersey — where no life can survive.

Fertilizers are a well-known and notorious source of water pollution but other environmentally destructive "additives" are more surprising — plastic bags and prescription pills. Plastic bags, manufactured by the trillions every year, require 1,000 years to decompose on land and 450 years to decompose in water. They are found in lakes and rivers the world over, where they clog wetlands and drainage systems and kill aquatic life. And prescription pills leak chemicals and hormones into our public water systems, affecting people for whom they were not meant. Chris Metcalfe, a water quality expert at Trent University in Peterborough, Ontario, says that 50 to 70 percent of all drugs pass through us. In water samples he has studied, he has found high levels of naproxen, used for both animals and humans as an anti-inflammatory, and carbamazepine, a drug once used for epilepsy and now prescribed for depression. Scientists in Germany have discovered compounds that make up such drugs as ASA, antidepressants, blood-pressure medications, ibuprofen, and beta blockers in the water supply of Germany and other European countries. Tests in Germany and Canada also found serious levels of estrogen from birth control pills in the local water supplies of the two countries.

LOSING THE GREAT LAKES

Originally formed from glacial meltwater beginning 12 to 20 millenniums ago, the Great Lakes contain 20 percent of the world's fresh water, making them the largest fresh water system on earth. So vast and deep are these lakes that only the top 75 centimeters (about 30 inches) — one percent of the total water volume — is renewed each year. Yet high levels of dioxins, polychlorinated biphenyls (PCBs), furans, mercury, lead, and scores of other noxious chemicals have been found in every lake at every depth. Most have appeared in the lakes in the last 50 years, reaching their waters from industrial and urban sources, via polluted groundwater, polluted overland flow, inflowing rivers and tributaries, and even the air.

Each year, between 50 and 100 million tons of hazardous waste is generated in the surrounding watershed, pesticides alone counting for 25 million tons. The International Joint Commission, which oversees the Canada-U.S. management of the lakes, reports that there are now serious build-ups of radioactive waste in the Great Lakes from the nuclear power industry. And the U.S. Environmental Protection Agency (EPA) has indicated that of roughly a hundred thousand sites around the lakes discharging industrial waste containing dangerous chemical substances, more than two thousand are directly contaminating groundwater.

Many of these toxins will never break down, and in a process called *bioaccumulation*, they move up the food chain in higher concentrations at every level. The total increase in concentration from source to humans at the top of the chain can be as high as a millionfold. Environment Canada says that a person who eats lake trout from Lake Michigan will be exposed to more PCBs in one meal than in a lifetime of drinking water from the lake.

Since the Great Lakes have been used as a common industrial dumping ground, less than 3 percent of the lakes' shorelines are now suitable for swimming, drinking, or supporting any aquatic life. The Nature Conservancy in the United States has identified 100 species and 31 ecological communities at risk within the Great Lakes system and notes that half don't exist anywhere else. And the pollution from the lakes is carried out into its tributaries, damaging the species living there. Among the creatures affected are the endangered beluga whales of the St. Lawrence River, which carry evidence in their bodies of toxic chemicals from the lakes.

The Great Lakes are also losing water, partly because of groundwater mining. Groundwater sources make up about half the volume entering the Great Lakes on the American side, and about 20 percent on the Canadian side. Fierce competition for this aquifer water is draining a crucial source for the lakes. And global warming is taking a terrible toll. Lake Superior's water level is at its lowest since 1926, and the once-extensive ice cover on the lakes has declined every year since the winter of 1993–94. Scientists at Great Lakes United, an environmental group with members from both Canada and the U.S., predict that if global warming continues at its current rate, the temperature of the Great Lakes will rise over the next hundred years by over 9°C (48°F), and lake levels will fall by a meter overall and by 2.5 meters (about 8 feet) in Lake Michigan. At current rates, in less than 40 years, the flow from the Great Lakes into the St. Lawrence River will have dropped by a quarter, potentially resulting in yet another great river of the world no longer reaching the sea.

The waters of the Great Lakes are also threatened by oil drilling on their shores. In Michigan, for instance, the state government plans to grant oil-drilling leases for 9 large wells on the shore of Lake Huron and 20 along Lake Michigan — in spite of warnings from environmentalists on both sides of the Canada-U.S. border of the disastrous effects these operations will have. Ontario has also quietly been issuing oil-drilling permits for smaller wells at about 20 a year since 1995, ignoring repeated predictions of future oil spills in the lakes.

Over time, the Great Lakes have been put at additional risk by the removal of the extensive wetlands that once edged their shores, acting as buffers against severe weather and protecting the shorelines from waves. These natural barriers have been mercilessly removed for industry and urbanization. Only 20 percent of the original wetlands still exist and these are decreasing by a staggering eight thousand hectares (almost twenty thousand acres) every year. Similarly, the once-extensive forests that blanketed the area have been largely cut down. Of the white pine forests that once covered almost half the region, only one percent remains. Their role in erosion control and purification of contaminants has not been replaced, and the Great Lakes are suffering for it.

WETLAND LOSS

Throughout North America, wetlands have acted as erosion control barriers and provided homes for fish and amphibians and resting grounds for migrating birds. They are an essential part of the habitat of 95 percent of all commercially harvested fish on the continent and a sanctuary for over half its endangered bird species. According to the Audubon Society, they are actually comparable to tropical rainforests in the diversity of species they support. Wetlands also act like sponges, soaking up excess rain and snow melt that would otherwise cause flooding, and they function like kidneys, filtering out dirt, pesticides, and fertilizers before the unwanted runoff reaches lakes and rivers. Once the water is purified, marshes and swamps serve as fresh water storage areas. And in strictly economic terms, each hectare of wetland is worth 58 times more than a hectare of ocean, since wetlands protect endangered species and commercially harvested fish.

Common sense would suggest that such a valuable resource should be nurtured and protected. Instead, about half the world's wetlands have been lost over the last century. In Asia, more than five thousand square kilometers (about 1,900 square miles) are destroyed every year to make way for industrial expansion, urbanization, and irrigation. In the United States, one hectare (about two and a half acres) is being lost every minute. In the continental U.S. as a whole, more than half the wetlands have disappeared. California has lost 95 percent, and fast-growing Florida has destroyed a wetland mass larger than all of Massachusetts, Delaware, and Rhode Island combined. As a result, populations of migratory birds and waterfowl have dropped from 60 million in 1950 to just three million today. And the wetlands that are the most biologically diverse are the most degraded, since they put the most species and wilderness at risk.

In Canada, water expert Jamie Linton has documented a disturbing story of water system abuse. In a study prepared for the Canadian Wildlife Federation, he observed that Atlantic Canada has lost 65 percent of its wetlands, southern Ontario has destroyed 70 percent, the Prairies have lost 71 percent, and in the Fraser River Delta in southern British Columbia, a staggering 80 percent has vanished forever. These stories are not unique. Wetlands cover only 14 percent of Canada's land area, and most have been lost to urban sprawl and large-scale farming.

DEFORESTATION

Forests also play a vital role in protecting and purifying sources of fresh water. They absorb pollutants before they run off into lakes and rivers, and like wetlands, they prevent flooding, particularly in southern countries subject to widely fluctuating cycles of drought and heavy rains. When forests are clearcut or depleted in nonsustainable ways, the integrity of local watersheds is threatened or destroyed, but when they are harvested wisely or left in their wild state, they can perform their functions as safety valves for rivers and their watersheds.

The Amazonian rainforest, famous for its diversity of animal and plant species, also acts as an ecological buffer for the Amazon River and the land around it. This great river flows 6,500 kilometers (about 4,000 miles) from the Andes to the Atlantic Ocean, contains one-fifth of the world's fresh water discharge into oceans, and creates habitat for 3,000 species of fish alone — more than any other river in the world. During the dry season, the forests around the river are relatively dry, but when the rains come (lasting about five to seven months every year), the river may rise to levels of up to 30 feet (about 9 meters). With no buffer, such a huge volume of water would wash the soil from the Amazon's banks into the river, leaving the land devastated. But the plants and trees of the Amazonian rainforest can provide erosion control because they are adapted to living submerged or half-submerged for a good part of the year. According to Brazilian climatologist Luiz Carlos Molion, these flooded forests intercept about 15 percent of the region's rainfall and act as a protective sponge, absorbing the huge volumes of seasonal rain. He says that removing these forests would lead to as much as 4,000 cubic meters per hectare (about 56,000 cubic feet per acre) per year directly hitting the ground and causing massive erosion of the soil into the Amazon River.

Although the destruction of this forest throws the whole system out of balance, deforestation is speeding up. In the lower third of the Amazon Basin, only 15 to 20 percent of the flooded forest remains as up to 17 million hectares (about 42 million acres) of tropical forest are destroyed each year. Of this amount, as many as 6 million hectares (about 15 million acres) are destroyed in Brazil alone, and the northern Brazilian states of Para and Maranhao have lost a forest area the size of Great Britain in just decades. Local authorities say the forests in the two states will be gone in a matter of years. In Chile, exploitation of the forest for export profit has exploded. Recent studies warn that the country will be depleted of its forests by 2025.

At the other end of the continent, Canada, which has almost 13 percent of the world's forest cover, is not doing much better. Logging is increasing with every passing year and the country now loses over one million hectares (about 2.5 million acres) annually, a hectare every three seconds. Moreover, most of this lumbering is done according to nonsustainable forest practices. As Elizabeth May of the Sierra Club of Canada reports, approximately 90 percent of logging in Canada's forests is clearcutting and about 90 percent of the area logged each year has not previously been commercially cut. So pristine areas continue to be devastated. When a forest is clearcut in a watershed, sudden influxes of sediment can destroy an aquatic ecosystem in minutes, blanketing the bed of a lake or stream and suffocating all the organisms living on the bottom. And the landslides that frequently follow clearcutting often contain pollutants that run directly into clean waterways.

In August 2001, the United Nations Environment Programme (UNEP) issued a dire warning to the citizens of the world in a report entitled *An Assessment of the Status of the World's Remaining Closed Forests*. In this study, the Environment Programme examined how many forests remain in the world with enough canopy to sustain watersheds and life. Only one-fifth of the planet is still covered with sustainable forests, stated the report, and few of those are protected by governments. Worse, the assault on those that are left is relentless. Klaus Toepfer, UNEP's executive director, was direct in his prognosis: "Short of a miraculous transformation in the attitude of people and governments, the Earth's remaining closed-canopy forests and associated biodiversity are destined to disappear in the coming decades."

GLOBAL WARMING

Most scientists in the world have come to a consensus on the phenomenon known as global warming, or climate change. For the last half-century or so, greenhouse gases — carbon dioxide, methane, nitrous oxide, and chlorofluorocarbons (CFCs) — have been discharged into the atmosphere in enormous quantities, with disastrous consequences. As we have denuded our planet of forests, thus heating up its surface, we have also been burdening the atmosphere with heat-trapping gases by burning fossil fuels. The result is predictable: a hotter planet.

According to the UN's Inter-governmental Panel on Climate Change, global average temperatures have risen 0.6°C (33°F) above the pre-industrial average. If emissions continue to rise according to current trends, greenhouse gas concentrations could reach double pre-industrial levels by 2080 — higher than they have been for several million years. This would cause a global average increase in temperature of 2.5°C (36.5°F), with possibly a 4°C (about 40°F) increase over land masses. These may seem to be small increases, but fourteen thousand years ago, when the earth's surface temperature increased by only four degrees, it was enough to end the Ice Age. Already oceans are rising because the polar ice caps are melting. Scientists point out that the warmest century of the last millennium was the 20th century; the warmest decade of the last millennium was the 1990s; the warmest year of that decade was the year 2000, and 2001 was warmer still. Not surprisingly, the oceans rose during the 20th century by about ten centimeters (about four inches), most of it occurring in the last half of the century.

As Simon Retallack and Peter Bunyard write in *The Ecologist*, "The implications [of global warming] for life are immense. With higher temperatures, there is more energy driving the Earth's climatic systems, which in turn causes more violent weather events. Severe storms, floods, droughts, dust storms, sea surges, crumbling coastlines, salt water intrusion of groundwater, failing crops, dying forests, the inundation of low-lying islands, and the spread of endemic diseases such as malaria, dengue fever and schistosomiasis is in the cards if the consumption of fossil fuels is not phased out . . . . Agriculture worldwide," they continue, "would face severe disruption and economies could tumble. There would be millions upon millions of environmental refugees — people fleeing from the intruding seas, or equally from the deserts they have left in their wake after stripping the land of its vegetation. Those are the prospects and scientific advisors to the UK government are warning that millions will die worldwide because of the processes of global warming that have already been unleashed."

An important part of this picture is the impact of global warming on fresh water sources. Wetlands, already at risk, will be adversely affected by growing droughts. According to the highly respected Hadley Centre, an environmental think tank in the United Kingdom, sea-level rise will result in about 40 to 50 percent of the world's coastal wetlands being lost by 2080. Directly under threat are the vast tidal mudflats, salt marshes, and sand dunes of the Netherlands, Germany, and Denmark, which are resting grounds for millions of migrating birds. Also at risk are the wetlands of the Mediterranean, the deltas of the Nile in Egypt, the Camargue in France, the Po in Italy, the Ebro in Spain, and at least thirteen thousand hectares (about thirty-two thousand acres) of English shoreline, much of it vital wildlife habitat. On other continents, climate change is predicted to lead to the disappearance of the mangrove forests of the west African coast, east Asia, Australia, and Papua New Guinea, which protect local lakes and rivers and act as breeding grounds for fresh water fish.

As global warming raises the surface temperature of the earth, the soil water needed to sustain the fresh water cycle evaporates more readily. Surface water (the water in lakes and rivers) also evaporates more and the snowpacks needed to replenish fresh water supplies become smaller and fewer. That is, when snow melts unseasonably, it evaporates instead of melting into the streams that feed lakes. And these lakes pose some problems of their own when they no longer freeze over. Water evaporates at a far slower rate when it is under ice cover, leaving more water behind to seep into the ground. When there is less freezing, more of that water is lost to the atmosphere. Similarly, as glaciers left over from the Ice Age melt, the river systems they feed will lose water. In Canada, the glacier that is feeding Alberta's Bow River is melting so quickly that in 50 years, there will not likely be any water left in the river except for the occasional flash flood.

Global warming also has a negative impact on the *residence time* of a lake. No water is static, but a given water molecule will stay in a particular area for a certain period of time. Residence time, as naturalist E.C. Pielou explains, is the average length of time that any particular water molecule remains in a lake; it is calculated by dividing the volume of water in the lake by the rate at which water leaves it. In northwestern Canada, climate change is already dramatically affecting the residence time of a number of lakes. In one study, rainfall in the area decreased by almost 1,000 to 650 millimeters (about 39 to 26 inches) every year, while above-average temperatures speeded up evaporation from the lakes. As a result, the residence time in just one of the lakes studied increased from 5 to 18 years over 15 years. This means that the lake is taking almost four times as long to renew itself as it did only several years ago.

Some scientists say that global warming is the single greatest cause of fresh water shortage in the world, and they predict the lowering of water tables in all the great lakes and rivers of the world. The Hadley Centre predicts that global warming will cause much of the Amazon Basin to become desert by 2050. And according to Dr. Nigel Arnell of Southampton

University in England, global warming alone will cause an additional 66 million people to be living in water-stressed countries by the year 2050 and 170 million more people to be living in countries that are severely water stressed.

INVASIVE SPECIES

Another potent threat to fresh water species and the lakes and rivers that sustain them is the introduction of non-native, or "exotic," species into an aquatic habitat. While not a result of pollution, they are a form of pollution, and they have become a greater threat as globalization and free world trade have increased the likelihood that non-native species will be transported to areas where they can do damage. As Janet Abramovitz of the Worldwatch Institute explains, exotics prey on native fish, compete with them for breeding space, and introduce new diseases. They can also mask ecosystem decline. An artificially stocked lake may have lots of fish, but if they are not part of the natural chain, the rest of the lake or river's wildlife may be in decline. Examples of exotic species invasions abound. Competition from non-native fish in Lake Victoria in Africa, for example, has driven 200 species of native cichlids to extinction and endangered 150 more. The introduction of the Nile perch for commercial fishing has virtually eliminated the native fish population.

The most documented case, however, concerns the Great Lakes of North America. Two hundred years ago, each of the five Great Lakes had its own thriving aquatic community. In 1900, 82 percent of the commercial catch was native. By 1966, native species were only two-tenths of 1 percent of the catch, and the remaining 99.8 percent were exotic species, most of them devastating to the local species. Abramovitz reports that some exotics, such as sport fish, were introduced, but that most came in by moving through canals constructed by humans or by hitchhiking on ships. The lamprey almost destroyed the annual catch of commercial lake trout in Lakes Michigan and Huron, and the zebra mussel, which was introduced from the Caspian Sea in 1988 by a ship's ballast, is now choking all of the major lakes and tributaries of the Great Lakes Basin, virtually eliminating the plankton that native fish and mussels need for survival. In fact, the

Great Lakes serve as an example of the effects of *all* the threats to fresh water that we've just described. They have sustained wetland loss, deforestation, invasive species, global warming, and massive toxic pollution. The result has been a catastrophic loss of biological diversity.

OVERIRRIGATION AND NONSUSTAINABLE FARMING

Where water resources have already thinned out, people have often turned to irrigation as a solution. This may appear to be a good decision, but the long-term effects of many kinds of irrigation are startling. Sandra Postel of the Global Water Policy Project has studied irrigation patterns all over the world and throughout history, and she has found that irrigation is not always beneficial. While it might feed billions of people and make the desert bloom, massive irrigation has within it the seeds of its own destruction. When water is used to grow food in an arid region, it causes the land to be overtilled. Then the soil breaks up into fine particles and blows away in the wind, leaving parched land and drought. The situation becomes worse when land is extensively irrigated without proper drainage. All water contains some salt, and irrigation water that is not properly drained leaves a salt residue. This salt then builds up, eventually making the soil unusable for farming. Combined with human-induced drought, salination has caused severe problems in China, India, Pakistan, Central Asia, and the United States. Salinity has affected a fifth of the world's agricultural land, and each year it forces farmers to abandon a million hectares (about 2.5 million acres) of farmland.

In her important 1999 book, *Pillar of Sand*, Sandra Postel has also traced the growth in irrigation use around the world. In 1800, global irrigated land totaled just 8 million hectares (about 20 million acres), but today, the irrigation base is 30 times larger. In the United States alone, the amount of land used for irrigation has doubled in the last 30 years, and humanity now derives 40 percent of its food from irrigated land. Aquifers also suffer depletion as a result of this form of farming. The Ogallala in the western United States waters one-fifth of the irrigated lands in the U.S. In China, irrigated areas grew by 2.5 percent in the last 50 years, and groundwater now provides close to 20 percent of the water needed for nearly 50 million hectares (about 124 million acres) of irrigated crops. To provide for this explosion in irrigated lands, China built over 2 million wells in the last 40 years. Around the world, there are now about 230 million hectares (about 570 million acres) of land under irrigation — up from just 6 million (about 15 million acres) two centuries ago. With these kinds of numbers, the pressure being placed on fresh water resources is self-evident. The big players — China, the United States, India, and Pakistan — account for more than half the irrigated land in the world, and all are experiencing increasing problems with drought, desertification, topsoil erosion, and water shortages.

In a June 2001 report, the United Nations Food and Agriculture Organization (FAO) said that one billion people live in arid countries where the land has been so damaged by overuse that it is incapable of producing enough food. Desertification now covers 3.6 billion hectares (8.9 billion acres) in more than one hundred countries, reports the FAO, and the situation is getting worse. In some extreme cases of overirrigation, whole water systems have dried up. Lake Chad, for instance, one of the last large water bodies in the center of Africa, has shrunk by more than 90 percent since 1960, and irrigation is acknowledged as the main culprit. The rapid retreat of the shallow lake and its depleting fish stocks could raise political tensions because the lake is shared by four countries. Lake Chad is bordered by Sudan, Chad, Nigeria, and Cameroon. It was once so large, it is thought it was a source of the Nile. But the lake's main sources of replenishment, the Chari and Logone rivers, were heavily taxed for irrigation starting in the early 1980s, and they have almost run dry. Adding to the crisis is the fact that the area is now in its third decade without its annual monsoon rains. This drought, along with the erosion of topsoil caused by intensive farming, has created disaster in an area where a lake and its surroundings once existed in healthy ecological balance.

The Zaindeh river in northern Iran has also dried up completely in 1999 because of poor irrigation practices. This river has served as the lifeblood for the people of Isfahan, in north-central Iran, and when it disappeared, it threw a hundred thousand farmers out of work and left ancient bridges and arches to oversee a dustbowl. Irrigation practices that turned a desert into a garden sowed their own seeds of destruction. Now the fertile farms *and* the water source are gone.

The acknowledged worst case, however, is the infamous Aral Sea, a basin shared by Afghanistan, Iran, and five countries of the former Soviet Union. An inland saltwater lake, it was once the world's fourth largest lake, fed by two powerful rivers, the Amu and the Syr. Years ago, the central planners of the former Soviet Union decided to irrigate the central plains of Asia and the deserts of Uzbek and Kazakh with water diverted from the rivers that feed the Aral Sea, in order to grow cotton for export. They created a vast system of mechanized agriculture based on intense irrigation and heavy use of pesticides and herbicides. For a while, the plan worked economically. Between 1940 and 1980, the Soviet Union became the world's second-largest cotton producer. But the experiment has been catastrophic for long-term prosperity, for the environment, and for the people of the region.

The Aral Sea has lost 80 percent of its volume and what is left is ten times saltier than it was. Its surrounding wetlands have shrunk by 85 percent. Almost all fish and waterfowl species have been decimated and the fisheries have collapsed entirely. Without the moderating influence of the sea, temperatures in the region have become more extreme and the season for growing food has shortened. Each year, winds pick up 40 to 150 million tons of a toxic salt-mixture from the dry sea bed and dump it on the surrounding farmlands. Millions of "ecological refugees" have fled the area. Those who stay face soaring cancer rates, partly as a result of heavy use of pesticides. In addition, a now-deserted island in the sea was once used for biological weapons testing and research by the Soviets. Because of the receding waters, the island will soon meet the mainland, and the germs and containments will be linked to the mainland. In a magazine article published in 1987, government water planners pronounced the sea nearly dead: "May the Aral Sea die in a beautiful manner," they wrote. "It is useless."

While the crisis has not reached such proportions in North America, drought, topsoil erosion, and water shortages are taking their toll. Peter Leavitt and Gemai Chen are biologists at the University of Regina in Saskatchewan. They have been studying sediment samples in lake beds to reconstruct the history of drought on the Canadian Prairies in order to better predict future droughts. Their conclusion: "The probability of the Prairies experiencing a severe drought in the next 30 years is very high." Dr. David Schindler, Canada's foremost fresh water scientist agrees. He says that due to global warming, the Canadian Prairies may become a dust bowl and Lake Manitoba may dry up.

Farmers and politicians in Canada are examining the possibility of mass irrigation to combat the coming problem, but they are being warned that this could lead to severe water shortages and the kinds of problems experienced on irrigated land in other parts of the world. For the prime examples, scientists point south to California, where people have spent billions of dollars diverting rivers some nine hundred kilometers (about 560 miles) away from their sources; to the American Midwest, where deserts bloom with water stolen from Nature; and to the Great Plains, America's breadbasket, where the land is being tilled to depletion.

The Great Plains once contained a vast natural storehouse of nutrients and minerals. The first plough turned the sod in the 1850s, and millions of years of stored wealth lured hungry people from all over the world to farm the rich land. But the practices used in more recent times — intensive farming and the heavy use of chemical fertilizers and pesticides — have been terribly harmful. Overtilling of the soil has exposed it to the elements; every acre is losing seven tons of topsoil every year. One-third of the topsoil and one-half of the nutrients that were originally found in the farmlands of the American Great Plains are gone.

Yet these very farming practices are encouraged by the American government, who give out US$28 billion a year in crop subsidies and tax breaks to farmers of this region who follow these damaging practices. The farmers can't resist; it's called "farming the government." The more bad practices they use, the more water they waste, the more money they get. In California, farmers rarely pay more than 20 percent of the real cost of irrigated water. So instead of planting crops that would be more appropriate to a semi-arid region, they grow crops like cotton. They also grow alfalfa, which is fed to beef cattle. It takes at least fifteen thousand tons of water to produce a ton of beef and nearly that much to produce a ton of cotton. To produce wheat or soybeans requires only 2 percent as much water. But the American government continues to subsidize these crops, paying farmers to waste water and erode the soil. And the story is the same almost everywhere in the world. The World Resources Institute reports that two-thirds of all agricultural land worldwide has been degraded in the last 50 years, and many practices that are damaging to farms also have disastrous effects on the world's fresh water systems.

DAMS AND RESERVOIRS

Many governments around the world have answered the increase in water demand by building more dams and diverting more rivers. The people of even the earliest civilizations, from the Romans to the Mayans, built aqueducts and set up irrigation schemes. The earliest recorded dam was built in Egypt 4,500 years ago and was made of earth, as all dams were until concrete was invented. But the natural channelization of water was changed forever when humans started to build permanent high-tech megastructures to harness the mighty flow of rivers.

In the 20th century, 800,000 small dams and 40,000 large dams (more than four stories high) were constructed, more than 100 of them behemoths over 150 meters (about 500 feet) tall. Of these, the vast majority have been built since 1950. The greatest numbers are found in China, followed by the United States, the former Soviet Union, Japan, and India. As a result of this construction, more than 60 percent of the world's rivers have been harnessed. In the United States, only 2 percent of the country's rivers and streams remain free-flowing and undeveloped; in Canada, more stream flows are diverted out of their basins of origin than in any other country in the world — by a considerable margin. Dam reservoirs have flooded about a million square kilometers (about 380,000 square miles) globally and hold a volume six times larger than the volume in all the world's rivers.

Dams are built for several reasons: to provide hydroelectricity; to facilitate navigation; to create reservoirs for cities and agricultural irrigation; and to control flooding. Once the symbol of human mastery over Nature, large dams have fallen into disrepute as evidence of their massive ecological impact steadily mounts. As Patrick McCully describes so thoroughly in his 1996 book, *Silenced Rivers*, the problem with reservoirs is that they require land to be flooded and submerged. The drowning of land vegetation creates the habitat required by the bacteria that absorb any mercury that happens to be in the soil. The reservoirs convert this mercury into a form that fish can ingest and mercury then enters the food chain. It bioaccumulates, and can be many times more lethal by the time humans eat it than in its original form. This is how the Cree of northern Quebec came to have such high levels of mercury in their systems. When they ate fish from the waters diverted for the massive James Bay hydroelectric project, 64 percent of the Cree in the area took in unsafe levels of this poisonous element. Mercury poisoning can cause blindness, reproductive failure, and brain damage.

Dam reservoirs also add to global warming as submerged and decomposing vegetation releases huge amounts of carbon dioxide and methane — two major greenhouse gases — into the atmosphere. A reservoir powering a hydroelectric plant can sometimes give off as great a quantity of greenhouse gases as a coal-fired generator. The most famous case of large-scale flooding of forest occurred in South America, reports McCully. Brokopondo Dam in Surinam submerged 1,500 square kilometers (about 576 square miles) of rainforest — one percent of the country. The decomposition of the organic matter in the shallow reservoir severely deoxygenated the water and caused such massive emissions of hydrogen sulfide, a corrosive and foul-smelling gas, that workers had to wear masks for two years after the reservoir started to fill in 1964.

Further, the tremendous weight of water in a basin not designed to hold it deforms the earth's crust beneath it, sometimes causing earthquakes. There is now documented evidence linking earth tremors to some 70 dams. In fact, the shift of weight when so much water is moved by human technology is affecting the earth's rotation. Geophysicists believe that dams have slightly altered the speed of the earth's rotation and the shape of its gravitational field.

Dams and reservoirs have an enormous impact on local ecosystems. Massive sediments bury riverbeds and block water channels, and they are the major reason why so many rivers no longer reach the ocean. Because they greatly multiply the surface area of water exposed to the sun, dams, especially in hot climates, can also lead to the evaporation of huge amounts of water. About 170 cubic kilometers (about 40 cubic miles) of water evaporate from the world's reservoirs every year, almost one-tenth of the total amount of fresh water consumed by all the world's major human activities. Consequently, salt is left behind in unnatural amounts, and this high salinity in many of the world's major rivers destroys wetlands and aquatic life and renders the surrounding soil unusable.

Fish are also greatly affected, particularly species like salmon, which migrate. When they come to a dam, they try to jump but can never pass over the obstruction, and many die in the attempt. Reduced river flows downstream from a dam kill fish habitat, as do warmer waters, deprived of oxygen. Before the Columbia River was dammed, 2 million fish returned every year to spawn; that number has now been cut in half. After the Pak Mun Dam was built in Thailand, all 150 fish species that had inhabited the Mun River virtually disappeared. After five years of research and study, scientists with the World Conservation Union reported to the UN-sponsored World Commission on Dams: "We conclude that . . . dam projects are the major cause of imperilment and loss of freshwater biodiversity."

Industrialized and nonindustrialized nations alike are caught in a web of toxic lakes, desertified farmland, and water-wasting practices that are threatening the lives and well-being of their citizens. In attempts to improve conditions by investing in projects like draining wetlands and building dams, well-meaning governments have built the foundations of a system that is turning on the very people it was meant to serve. But now that we are aware of the damaging, and even catastrophic, effects of many of these practices, there is no reason to continue along the same path. Unfortunately, in addition to the inertia that can take over when people are confronted with the need to change, blind and ill-intentioned governments and corporate greed are combining forces and accelerating the pace at which water is being poisoned and lost. In the end, governments and corporations will pay a heavy price, but in the meantime, many private citizens are suffering.

DYING OF THIRST

*How the global water crisis
threatens humanity*

T he world's water crisis is having a devastating impact on quality of life
for billions of the world's citizens caught between the twin realities of
water scarcity and water pollution. In fact, the world's intensifying water
crisis is literally the arbiter of life and death for a growing number of
people. It is also becoming a matter of fierce competition and struggle
within societies and social classes and between nations.

The 3,400-kilometer *maquiladora*, or export-processing zones, on the
border between Mexico and the United States are toxic cesspools. Rivers
and streams in the region are so polluted that only 12 percent of the
people living there have reliable access to clean water, and many residents
live in homes with no sewage systems whatsoever. In the shantytowns
and cardboard shacks that surround the free trade zones, where precious
drinking water is delivered by truck once a week, the absence of fresh water
has become a symbol of the poverty of the more than one million people
who have flocked to the area in the last five years. And the filthy water
of the region carries disease and causes severe diarrhea. Though residents

drink the trucked-in water, they use local water for cooking and bathing and for irrigating crops, which then become dangerous to eat.

The filth of the *maquiladora*, its lethal water, and its squalid poverty push thousands of young Mexicans away from their home country. Every night, they head for the border, trying to enter the United States illegally to seek a better life. The strips at the crossings are notoriously dirty and dangerous. Six-lane highways divide cities like Tijuana and Juárez from the desolate belts of dirt where the men gather at dusk. A steep cement drop leads to a slow-moving river of chemical sludge and raw sewage about two feet deep. On the other side, the cement wall inclines at a 90-degree angle, fenced in at the top by a huge barbed-wire electric fence and lit by floodlights. The stench along the strip is unbearable; human and animal excrement, used condoms and needles, and piles of garbage mix with the stinking little river of sludge the men must run through to get to the other side. The chemicals and sewage get on their feet and into their shoes. Whether they reach the U.S. or are caught by American security patrols and sent back, they have had to pass through this deadly, filthy river and may have to do so again.

Also at the crossing are the people who can find no work in the *maquiladora*. They hang around the border at night, peddling fried tacos, condoms, and drugs — and plastic bags. Even the poorest of the would-be illegal immigrants will shell out the little they have to wrap plastic bags around their feet, for protection from the poisoned water.

LETHAL WATERS

Half the people on this planet lack basic sanitation services. Every time they take a drink of water, they are ingesting what Anne Platt of the Worldwatch Institute calls water-borne killers. So it is not surprising that 80 percent of all disease in the poor countries of the South is spread by consuming unsafe water. The statistics are sobering: 90 percent of the Third World's wastewater is still discharged untreated into local rivers and streams; water-borne pathogens and pollution kill 25 million people every year; every eight seconds, a child dies from drinking contaminated water; and every year, diarrhea kills nearly three million children, a full quarter of the deaths in this age group. The declining quality of the world's water has also caused malaria, cholera, and typhoid to occur more frequently in many places where they had been all but wiped out. They proliferate in conditions of dense population, poor sanitation, and poverty. Between 1990 and 1992, the number of people suffering from cholera worldwide rose from one hundred thousand to six hundred thousand, and these numbers continued to rise throughout the decade, though not so sharply.

In 1991, a pollution crisis caused a particularly bad cholera outbreak. In that year, a Chinese ship dumped its sewage into a bay in Lima, Peru, and within three weeks, cholera had spread up and down the coast, causing acute diarrhea, severe dehydration, and sometimes death. In the first year alone, nearly three thousand Peruvians died. Over the next two years, this one outbreak gradually contaminated the water supply of all but two countries in Latin America, infecting five hundred thousand people.

People living in Africa are afflicted by many different water-borne diseases. As many as two hundred thousand are thought to suffer from schistosomiasis, or bilharzia, a disease borne by water snails often found in irrigated water taken from dams. It causes cirrhosis of the liver and intestinal damage. Some 18 million Africans have onchocerciasis, or river blindness, carried by a blackfly that breeds in dirty rivers. And during the Sudan civil war of 1997, thousands of people fleeing from the war drank putrid water in refugee camps and became infected with sleeping sickness, a brutal illness carried by the water-bred tsetse fly.

Some disease-causing organisms — *cryptosporidium, E. coli, giardia* — are directly linked to poor or nonexistent sewage treatment and are also making a comeback. They are caused by too much human and animal sewage leaking into drinking water. In some cases, these afflictions are a result of too many people living in too little space, right near their sources of untreated drinking water. A family in Africa may have to defecate in a place not far from their well, and if they have livestock, the animals may have to do the same. A child in a shantytown in the Philippines might have no choice but to defecate near the family's water tank, and whole villages may drink water from a river that also carries raw sewage.

In other cases, government cutbacks are affecting the quality of drinking water. In Canada, the Ontario provincial government massively cut the budget of the Ministry of the Environment, gutted the water protection infrastructure, and laid off many trained water-testing experts. Some of the testing previously done by government workers was farmed out to private testing labs. Then, in 1999, a Canadian federal government study revealed that a third of Ontario's rural wells were contaminated with *E. coli*, and in June 2000 at least seven people, one of them a baby, died from drinking the water in the little town of Walkerton.

Back in Africa, mounting debt repayments during the 1980s and 1990s forced many countries to cut down on water and sanitation services to their citizens. These are only some of the Third World countries that are still paying as much as 70 percent of their national budgets to repay their debts to the International Monetary Fund and the World Bank. The tragedy that ensued was documented by Peter Gleick of the Pacific Institute for Studies in Development, Environment, and Security, a highly respected think tank on water issues based in California. In Nairobi, capital expenditures for water fell by a factor of ten in just five years during the 1980s. In Zimbabwe, 25 percent of village water pumps failed when the government cut maintenance funds by more than half. Dysentery rates in Kinshasha soared in 1995 when funds for water chlorination ran out, and cholera cases and deaths rose dramatically.

⌒

In South Africa, a recent cholera outbreak has been directly linked to the government's decision to cut off water supplies to those who could not pay their water bills. More than 100,000 people in KwaZulu-Natal province became ill with cholera, and 220 died in the course of ten months beginning in August 2000, after the South African government, urged by the World Bank, implemented a "cost recovery" program and denied water and sanitation services to many thousands of citizens who had been getting their water free of charge.

Some diseases are linked to modern pollution and afflict people living in the industrialized nations of the North as much as or more than those living in the Third World. Lead has been linked to loss of intelligence and behavioral problems in children in both the South and the North. Sixty billion people on the Indian subcontinent have been poisoned by fluoride.

The combination of household and industrial disinfection by-products, along with chlorine added to water, has been implicated in cancer deaths. Arsenic has been associated with bladder, skin, and lung cancers. Over the last decade, high levels of this poison were found in Bangladesh, where one in five water pumps was contaminated with high levels of arsenic. Although in these cases, the arsenic occurred naturally, rather than being the result of toxic dumping, people were forced to dig deep into wells because of severe groundwater shortages and pollution. If they had not had to dig so deeply, they could have avoided the arsenic.

The U.S. Environmental Protection Agency (EPA) estimates that more than half the wells in the United States are contaminated with pesticides and nitrates. Pesticides and chemicals such as perchloroethylene, or "perc"; PCBs; and dioxins accumulate in the body fat of animals, fish, and humans and are linked to cancer. In infants, reports the U.S. group Physicians for Social Responsibility, ingestion of high concentrations of nitrates from well water results in methemoglobinemia, which carries an 8 percent fatality rate. The Women's Environmental Network in Britain claims that as many as 8 percent of the country's children have sustained some nervous-system damage and memory loss as a result of exposure to dioxins and PCBs, and the World Health Organization has said that the increasing use of pesticides is killing forty thousand people every year.

Some diseases are even related to poor maintenance of water infrastructure. Half the people in the industrialized nations of the North, and more in poor countries, carry the stomach bacterium *helicobacter pylori*, usually caused by slime build-up in water pipes. The bacterium causes stomach ulcers and cancer and is particularly prevalent in unchlorinated well water and water supplies in Third World countries.

UNEQUAL ACCESS

There is nowhere on earth to go to escape the global water crisis. For instance, in the wealthy United States, says the Natural Resources Defense Council, some 53 million Americans — nearly one-fifth of the population — drink tap water contaminated with lead, fecal bacteria, or other serious pollutants. Alarmingly, according to the Environmental Protection

Agency, outbreaks associated with groundwater sources grew by almost 30 percent between 1995 and 1998 in the United States.

However, it is also very clear that the world's poor are taking the brunt of the crisis, whether we are talking about water-borne diseases or outright scarcity. A report from the United Nations Economic and Social Council to the UN's Commission on Sustainable Development says that fully three-quarters of the population living under conditions of water stress — amounting to 26 percent of the total world population — are located in Third World countries. By 2025, the commission projects, the citizens of low-income countries experiencing water stress will amount to 47 percent of the total world population. Furthermore, the great majority of the megalopolises in which more than 50 percent of the population has no access to clean water are located in the Third World, and the highest rate of growth within these cities is in the slums. By 2030, says the UN, more than half the population of these huge urban centers will be slum dwellers with no access to water or sanitation services whatsoever.

It is often said that the population explosion is a "water bomb" about to be detonated. There is no question that there is truth in this concern. Every year the world population grows by another 80 million people who have to share dwindling fresh water supplies. But as water expert Riccardo Petrella points out, that argument alone may encourage some to lay all the blame on the Third World, where the majority of the population explosion is taking place, and to overlook the fact that people living in the countries of the North consume much more water — among other goods — than people living in Third World countries.

The richest fifth of the world accounts for 86 percent of consumption of all goods. As Petrella explains, a newborn baby in the West, or a rich one in the South, consumes between 40 and 70 times more water, on average, than one in the South who has no access to water. North Americans use 1,280 cubic meters (about 45,000 cubic feet) of water per person every year; Europeans use 694; Asians use 535; South Americans use 311; and Africans use 186. Although the average European consumes only about half as much water as the average North American, their consumption levels are still high compared to citizens of nonindustrialized countries. And ironically, as the United Nations reports, Europeans spend US$11 billion a year on ice cream — US$2 billion more than the estimated amount needed to provide clean water and safe sewers for the world's population.

The disparity in consumption levels of North and South is partly a reflection of the fact that some parts of the planet have more fresh water supplies than others, but this is not the full explanation. For instance, Australians, who occupy the driest land mass on earth, use 694 cubic meters (about 24,500 cubic feet) per year — the same amount as Europeans — because their consumer-based culture results in high volumes of water wastage. Conversely, China has almost as much fresh water as Canada, but because of population demands and pollution of its surface waters, that country is considered an area of crisis.

The countries of the North are responsible for a disproportionate amount of the world's water consumption, partly because of individual habits and lifestyles. Citizens of the most privileged countries simply take water for granted or are able to buy it, even if it is expensive. And their lifestyles — SUVs, lawns and golf courses that are watered and sprinklered, swimming pools, and toilets that consume 18 liters (about 4.8 US gallons) of water per flush — use vast amounts of water. Another important factor in the consumption disparity is the use of water by industry. While globalization is spreading industrialization all over the planet, most industries are still located in the North. And where there is industry, water consumption is rampant. While agriculture still accounts for the vast majority of water use in the Third World, industry uses as much water as agriculture in North America and almost twice as much as agriculture in Europe. The water resources of the so-called Developed Nations are not yet as scarce as those of the Third World, but they are being wasted by a lifestyle based on water-depleting consumerism.

~~~

North Americans and Europeans have set themselves on a path that is leading to water scarcity. So far, resources in these nations appear to be abundant, but they are not infinite, and current rates of consumption will lead to depletion — especially as nonindustrialized nations try to emulate North American lifestyles. If these trends continue, we will, in time, be living on a water-scarce planet. To get a glimpse of what that future will look like, we can go to the Third World today. In crowded Asian, African, and Latin American countries, massive increases in animal and human waste, intensified by the establishment of factory farms, are exposing more and more people to cholera and deadly diseases caused by the *E. coli* bacterium. Most local governments cannot even afford basic chlorine to treat their water. And where local communities once turned to aquifers and hand pumps to avoid the problem of polluted surface water, chemical and human waste seeping into these sources has now made groundwater dangerous as well. In China, 80 percent of the population drinks contaminated water. In Papua New Guinea, one-quarter of the inhabitants live in critical conditions because they lack clean water, even though the country is water rich. And in India, 70 percent of the population has no proper drainage system. In Manila, in the Philippines, water shortages affect 40 percent of the residents, and in most Third World cities, water is often rationed to neighborhoods for only a few hours a day or a few days a week.

The people of Africa have suffered more than many as a result of bad water. Of the 25 countries that the UN lists as having the least access to safe water, 19 are in Africa, and people living on that continent have the highest rate of death from diarrhea, as well as a high incidence of malaria and other water-related diseases. Water in Nairobi is so scarce that slum dwellers have started to tap into wastewater mains, and for about 15 million South Africans, the nearest source of water is at least one kilometer away. According to Water Policy International, South African women collectively walk the equivalent of going to the moon and back 16 times a day just to fetch water.

ELITE PRIVILEGE

Although water inequality parallels the inequities that exist between the industrialized and nonindustrialized countries of the world, disparities also occur within individual societies. Surprisingly, the poorest people in poor countries pay much more for their water than the rich of their society. Municipal water, subsidized by governments, is delivered to the wealthy, and people in the middle class can install a small water tank for trucked-in water or dig a well. (The richer among the middle class can afford to sink deeper boreholes when the well water starts to dry up.) But the poor buy water by the can from private water carriers who may charge as much as one hundred times the rate of municipally delivered water. Anne Platt of Worldwatch Institute reports that a family in the top fifth income groups in Peru, the Dominican Republic, or Ghana is, respectively, three, six, or twelve times more likely to have water connected to their home by pipe than a family in the bottom fifth in those countries. Because they lack access to publicly subsidized utilities, says Platt, the poor often end up paying more for their water than do the rich because they must obtain it from illegal sources or private vendors.

In Lima, Peru, for instance, poor people may pay a private vendor as much as three dollars for a cubic meter of water, which they must then collect by bucket and which is often contaminated. The more affluent, on the other hand, pay 30 cents per cubic meter (about 35 cubic feet) for treated water provided through the taps in their houses. Hillside slum dwellers in Tegucigalpa, the capital of Honduras, pay substantially more for water supplied by private tankers than they would even if they paid for the government to install a water pipe. In Dhaka, Bangladesh, squatters pay water rates that are 12 times higher than what the local utility charges. And in Lusaka, Zambia, low-income families spend, on average, half their household income on water.

The elite of a nation and wealthy tourists also have special water access. In 1994, when Indonesia was hit with a major drought, residents' wells ran dry, but Jakarta's golf courses, which cater to wealthy tourists, continued to receive one thousand cubic meters (about 35,000 cubic feet) per course per day. In 1998, in the midst of a three-year drought that dried up river systems and further depleted aquifers, the government of Cyprus cut the water supply to farmers by 50 percent while guaranteeing the country's two million tourists a year all the water they needed. And where race and class come together, water privilege can be startling. In South Africa, six hundred thousand white farmers consume 60 percent of the country's water supplies for irrigation, while 15 million blacks have no direct access to water.

In Mexico, the situation is not much better. In the *maquiladora* zones near the U.S. border, clean water is so scarce that babies and children drink Coke and Pepsi instead. And during a drought crisis in the northern part of the country in 1995, the government cut water supplies to local farmers while ensuring emergency supplies to the mostly foreign-controlled industries of the region.

FOOD SCARCITY

As societies all over the world have become more dependent on irrigated land to grow their food, the lack of fresh water sources also threatens their food supplies. Simply put, many of the world's most important food-producing regions are running out of water for irrigation. As mentioned in Chapter 2, humans obtain 40 percent of their food from irrigated land, and the amount of irrigated land has grown exponentially in the last several decades.

This change in the basis of humanity's food production has put a profound strain on the world's groundwater supplies. The global harvest of fruit, vegetable, and grain crops uses an enormous quantity of water — so much that author Sandra Postel of the Global Water Policy Project in Amherst, Massachusetts, notes in her book *Pillar of Sand* that many important food-producing regions are sustained by the hydrological equivalent of deficit financing. As irrigators draw on water reserves to support current production, they are racking up large water deficits that will one day have to be balanced. Annual depletion (that is, net water loss) in India, China, the United States, North Africa, and the Arabian Peninsula alone adds up to about 160 billion cubic meters (about 5,650 billion cubic feet) a year.

Postel estimates that about 180 million tons of grain — approximately 10 percent of the global harvest — is being produced by using water supplies that are not being replenished. To feed the world's population by 2025, an additional two thousand cubic kilometers (about 476 cubic miles) of irrigation water will be needed. But since agricultural operations are already creating water deficits, Postel points out, where are farmers going to find the additional irrigation water needed to satisfy the food demands of the more than two billion people expected to join humanity's ranks in the next several decades?

DAM FALLOUT

The human suffering caused by big dam projects (linked, of course, to the massive increase in irrigation practices around the world) is as serious as their environmental fallout. An estimated 60 to 80 million people have been displaced by the building of dams around the world in the last six decades. These legions of "oustees," as they are called in India, have been culturally, economically, and emotionally devastated by the loss of community, livelihood, and links to their ancestral homes. This situation is familiar to the International Rivers Network, a U.S.-based group that was instrumental in convincing the United Nations to set up a commission investigating dams. According to Patrick McCully of this organization, families have often been flooded out with minimal or no compensation, and millions of independent farm families have ended up as slum dwellers on the edges of the Third World's burgeoning cities. These numbers do not even take into account the untold millions who are also negatively affected by massive diversion projects, but who still live on the land or rivers nearby.

India and China have created the largest number of oustees and have used brutal tactics to enforce their evictions. As an average of more than six hundred large dams were built every year in the three decades following the Chinese Revolution of 1949, at least ten million people were displaced, according to the Chinese government. But other observers put the number much higher. Chinese dam critic Dai Qing puts the total at more than 40 million. And many of the evictions have been carried out brutally. In 1958, for instance, hundreds of thousands of people were evicted to make way for China's Xinanjiang Dam. Officials, who ordered that the resettlement be carried out "like a battle action," sent in laborers to tear down houses, and traumatized peasants were forced to walk for days to resettlement sites. And more recently, nearly two hundred thousand people have been displaced for the massive US$4billion Xiaolangdi Dam now being built on the Yellow River. Observers fear a repeat of the failure of the Sanmenxia Dam built upstream on the same river in the 1950s. That dam deposited massive amounts of sediment and flooded riverbanks. Mao ordered it to be bombed when it threatened the ancient city of Xian. It had to be redesigned, and this fortified construction flooded 66,000 hectares (about 163,000 acres) of fertile farmland. According to a World Bank report, the majority of the 410,000 people displaced by the Sanmenxia Dam still live in abject poverty, without any means of livelihood.

Violence and intimidation face those objecting to evacuation to make way for the internationally contentious US$50 billion Three Gorges Dam. Probe International reports that in August 2000, residents living in a local resettlement village were beaten and put under house arrest by soldiers because they were voicing peaceful protests against the submergence of their ancestral homeland. Altogether, the Three Gorges Dam will displace 1,100,000 people, who may also face brutal methods of resettlement. One villager displaced by the dam had this to say to Probe International: "Officials can decide whether people live or die. Without the survival rights, we dare not tell anyone our real names. We will face terrible disasters if the county government knows we talk to you. We are being watched by people assigned by the county government. If found out, the outsiders [environmentalists and journalists] will be beaten up first and then inspected."

According to the World Commission on Dams, between 16 and 38 million people have been displaced by dams in post-Independence India. In 1981, for instance, one hundred thousand people living in the submergence zone of the Srisailam Dam in Andhra Pradesh were driven out of their homes in what authorities called "Operation Demolition." And over four hundred thousand have been directly evicted in the construction of the Sardar Sarovar (formerly the Narmada), now stopped by a court injunction. Like China, India has taken a brutal approach to dam evictions. In 1961, the Indian finance minister spoke to local farmers at a meeting in the submergence zone of the Pong Dam. He was blunt: "We will request you to move from your houses after the dam comes up. If you move, it will be good, otherwise we shall release the waters and drown you all."

Other evictions around the world have included acts of unspeakable barbarity. In the Soviet Union, evictees were often forced to take part in the burning and destruction of their own houses, orchards, and churches and the exhuming of relatives' coffins. But even worse was a case cited by Patrick McCully of International Rivers Network involving the murder of 378 Maya Achi Aboriginal people in Guatemala. In the early 1980s, a

European consortium threatened to displace 3,400 people in order to build Guatemala's Chixoy Dam. In spite of the obvious presence of thousands of residents, the consortium's feasibility study claimed that the land held "almost no population." The Maya Achi from the village of Rio Negro objected and asked for a fairer resettlement payment. But instead of receiving better compensation, on three separate and horrific occasions, Guatemalan soldiers appeared and massacred the Maya Achi men, women, and children.

This story is symbolic of how Indigenous peoples have traditionally been shoved aside to make way for dam projects. All over the world, their livelihoods have been disproportionately affected. In India, 40 percent of all those who have been displaced by dams are *adivasis* — low-caste Indigenous peoples who represent just 6 percent of the population. Almost all the large dams built in the Philippines are on land where Indigenous peoples live. In 1948, the Garrison Dam in the U.S. flooded most of the North Dakota Native reservation and displaced the majority of the people living there. More recently, the Innu peoples of northern Quebec suffered loss of habitat and traditional fish-spawning grounds when rivers near James Bay were inundated as part of a hydro project in northern Quebec in the 1970s.

Dams create the habitat for the parasites that cause schistosomiasis and other water-borne diseases, and the victims are often the oustees or people living downstream from the reservoir. After the construction of the High Aswan Dam in Egypt, schistosomiasis became endemic. After the Akosombo Dam in Ghana was flooded in 1964, displacing eighty-four thousand people, 90 percent of the children living near the reservoir became afflicted with the disease. Similarly, reports Patrick McCully, onchocerciasis, or river blindness, is carried by the irrigated waters from dams. So is malaria, a disease that is on the rise again after intensive eradication programs. Malaria breeds in the billions of mosquitoes found in stagnant water in hot, humid climates, but the ecological changes caused by dams and irrigation projects in arid and semi-arid areas also create the ideal breeding grounds for the disease. Malaria kills over one million people worldwide every year. In the area of Sri Lanka's five-dam Mahaweli project, stagnant water pools became a breeding ground for malaria-carrying mosquitoes, and in 1986, the first-ever case of malaria was reported in the region. Similarly, an epidemic of malaria, a disease which had been eradicated in southern Brazil, followed the construction of the Itaipú Dam in 1989.

WATER CONFLICTS

Given the reality of shrinking fresh water supplies, the pollution of existing sources, and the growing demand for water, it is inevitable that conflicts will arise over access. All over the world, communities in water-stressed countries are beginning to compete with one another for prior use of this precious resource. Tensions are growing across nation-state borders and between cities and rural communities, ethnic groups and tribes, industrialized and nonindustrialized nations, people and Nature, corporations and citizens, and different socio-economic classes.

Urbanization is also adding pressure to the already uneasy situation. As people move or are displaced to growing urban centers, demand for water will also rise in those places. So water is being diverted from rural and wilderness areas to meet urban demand, but farmers, already stretched to feed a burgeoning population, are understandably reluctant to let those precious water allocations go. As described in Chapter 1, this transfer is well under way in China, where the urban migration is just beginning in earnest. Cities and industry are favored openly by the Chinese government, and farmers are having their water taken from them without consent or even prior knowledge. The same situation is occurring in India. Some farmers are making more money selling their groundwater to urban and industrial users than they once made from growing food.

In *Pillar of Sand*, Sandra Postel reports that rice farmers in parts of the Indonesian island of Java are losing water supplies to textile factories, even though the law requires priority to be given to agriculture. To make matters worse, the factories sometimes take more than their permits allow, leaving farmers high and dry. They have also polluted local water supplies, which has a damaging effect on crops. In South Korea, farmers south of Seoul recently armed themselves with hoes and blocked municipal water trucks from pumping water for city dwellers, for fear of having too little water left for their crops. And in the American Pacific Northwest, farmers in the Columbia River Basin were paid forty dollars an acre in the summer of 2001 not to irrigate their crops, so the massive hydroelectric generators along the river could supply power to California.

In some cases, rural dwellers are pitted against other rural dwellers. In Brazil's Northeast, prolonged drought is fomenting strife between water haves and have-nots. The powerful São Francisco River has been diverted for irrigation, and what remains of it now snakes its way through what was formerly some of Brazil's bleakest terrain. As reported by Joelle Diderich for Reuters News Agency, this irrigation program has transformed 300,000 hectares (about 740,000 acres) of the dry river valley into orchards growing tropical fruits such as coconuts and guava for export. The state-run project is also financing roads, sewage systems, and an airport. Some might defend this gargantuan enterprise by pointing out that it has become a magnet for farmers seeking work and has created a wealthy few. The handful of farmers who came to the project first have become prosperous, but there are a limited number of lots, and farm workers have no job security. This has heightened inequities within the region. Drought also threatens more than 10 million people — the vast majority of the area — with starvation. This reflects the situation throughout Central America, where drought endangers more than half the 35 million citizens — some of it caused when huge quantities of water are used up by large, corporate-run, farm-for-export operations, leaving little for local family farmers to raise their crops.

Water scarcity has also pitted farmers against Indigenous peoples and those defending endangered species. In Klamath Falls, Oregon, for example, during the long, hot summer of 2001, farmers took the law into their own hands and repeatedly reopened reservoir gates for irrigation water that the Federal Bureau of Investigation had ordered closed, to protect endangered bottom-feeding suckerfish and threatened coho salmon during a year of record drought. Local Native tribes also had treaty rights to these fish and had demanded government protection of the fish and for water to be diverted back to their natural habitat. The Native people and commercial salmon fishermen farther downstream say that favoring farmers with massive irrigation for decades has deprived them of their livelihood and cultural rights.

For the farmers, however, the order was devastating. It cut off water to about 1,400 family farmers and ranchers in the Klamath Basin, and about two hundred thousand acres (about eighty thousand hectares) that grow water-thirsty alfalfa — normally kept green through irrigation — are now parched. The local community was so supportive of the farmers that the sheriff wouldn't arrest the trespassers who reopened the reservoir gates, and local prosecutors refused to prosecute. In a similar dispute involving farmers in California, a precedent-setting federal claims court decision determined that the redirection of water supplies to help endangered species in that state in the early 1990s constituted a taking of property and ordered compensation to the farmers.

Once more, the demands of the so-called free market placed farmers in the awkward position of entering into large-scale, water-hungry enterprises in order to increase volume of production and make up for the paltry prices they are paid for their crops. Once invested in highly mechanized, large-scale operations, farmers cannot continue to operate without depending on massive use of resources such as fossil fuels and water. Ironically, they are then placed in a position where they can do considerable damage to ecosystems and contravene Native rights. If farmers were encouraged and enabled to switch to more drought-resistant crops and less fuel-intensive farming, water conflicts like the ones just described would be less frequent. But for this to happen, the American government — like other governments around the world — will have to stop subsidizing industrial, resource-depleting agriculture and support sustainable, smaller-scale farming and encourage the cultivation of more drought-resistant crops.

NATURE AND POWER

Unemployed people can also be used as pawns in schemes that do damage to Nature. In Canada's easternmost province, Newfoundland, unemployment is chronically high, but many wilderness areas still abound with life. On the province's south shore, for instance, is a lake full of pristine water, 16 kilometers long by 10 kilometers wide (about 10 by 6 miles). Starting in 1997, a local businessman has applied to the province for an export licence, so he can sell Gisborne Lake water to thirsty consumers around the world. Not surprisingly, the proposal is extremely contentious. On one side are many Canadians concerned about the ecological effects of massive water removals and about losing control of the water supplies Canada needs to support its growing population — including future immigrants from water-starved countries. These concerns are heightened by the fact that Canada has short-sightedly signed trade agreements in which water is described as a tradable commodity. On the other side of the argument are the residents of Grand Le Pierre, a small fishing community of 350 located near Gisborne Lake, which was impoverished when the cod they depended on were overfished almost to extinction. With unemployment rampant, the community leapt at the opportunity to create jobs associated with the proposed project. The former premier, Brian Tobin, closed the door on water exports, but the new premier, Roger Grimes, has reopened it and the issue is being hotly contested.

Water disputes have also broken out between small farmers and the interests of agribusiness. In Ecuador, for instance, a new law on water is currently on the table and two opposing agricultural groups are working to shape it in quite different ways. As reported by water expert Riccardo Petrella, one of these proposals has been advanced by the Agricultural Chambers of Commerce and defends the interests of the big farmer and agribusiness. They tend to support privatization of water services and want water to be used for greater industrial productivity. The other proposal is offered by CONAIE (the Ecuadorian Indigenous Nationalities Confederation) on behalf of small farmers and workers. This statement maintains that water is a public asset that should be used mainly to serve the equitable development of the country's entire population. CONAIE contends that the food and water security of the local population should be the number one priority.

In many countries in the world, the elite of society are gaining privileged access to water, to greater and greater degrees. Throughout southeast Asia, for instance, golf tourism is on the rise, but this is creating a strong backlash from local residents who believe that the water-guzzling courses are "water-favored" by governments because of the tourism dollars they bring in. Nevertheless, golf courses continue to spring up all over this part of the world: Malaysia, Thailand, Indonesia, South Korea, and the Philippines maintain 550 golf courses, and they are building another 530.

The disparity of water access between rich and poor was violently dramatized in Bangladesh in 1999, when hundreds of residents of the capital city, Dhaka, attacked a power supply office, barricaded roads, and burned vehicles in the spring of that year to protest the scarcity of running water. The Dhaka Water and Sewage Authority admits that more than 30 percent of the city's nine million residents have no access to drinking water. The residents contend that the poor have been left by governments to fend for themselves.

Water strife can also be based on historic struggles of racism and power. Under apartheid, South Africa was openly discriminatory in its distribution of water. So the country's first democratic government inherited a serious set of water problems: water scarcity, unequal distribution of water based on race and class, severe pollution of water sources, heavily dammed rivers, and substandard or nonexistent sanitation for the black majority. At first it seemed that the new government understood these deep-seated social inequities and were prepared to eliminate water discrimination. In fact, South Africa's majority party set out to remedy the inequality by guaranteeing every person basic water rights in the new constitution. The African National Congress's Reconstruction and Development Programme declared that household water access as a human right was the "fundamental principle of our water resources policy."

However, in a study of post-apartheid water distribution, water scholars Patrick Bond and Greg Ruiters found that the African National Congress had also adopted a market-oriented approach to water management, thereby building in continued water shortages for the poor majority and water privilege for those who could pay. The government emphasized that water was a "scarce" good, requiring marginal cost pricing, even for the poor. Bond and Ruiters also discovered that the new government retained the apartheid-era hydrological bias toward supply enhancement through building expensive dams, while failing to charge a sufficient price to those who were wasting water. "The result was drought for those most in need of water and excess liquidity for those most prone to abuse it." And worse, water access and sanitation services to the majority of South Africans had actually declined during the first half-decade of democracy. A lower percentage of South Africans now enjoyed access to affordable water in their homes and yards than they had in 1994, and hundreds of thousands of water consumers had had their taps shut off in the late 1990s.

As a result, the distribution of South Africa's water throughout the population is even more inequitable, measured in class, race, and gender terms, than the distribution of income. More than half of South Africa's raw water is used for white-dominated commercial agriculture, and half of that water is wasted in poor irrigation practices. Another quarter is used in mining and industry. About 12 percent of South Africa's water is consumed by households, but of that amount, more than half goes into white households, including water for gardens and swimming pools. And, as stated earlier, 16 million South African women still have to travel by foot at least one kilometer to supply their families with basic water needs. In September 2001, police shot 15 people, one a five-year-old child, when residents resisted the cutting off of water to 1,800 people in Cape Town's Unicity. Private workers and protection services, guarded by hundreds of police, succeeded in cutting the water lines, leaving behind a devastated community. Fires broke out as local people, weeping openly, set up burning barricades to prevent retaliation.

BORDER STRUGGLES

About 40 percent of the world's population relies on the 214 major river systems shared by two or more countries. As water travels from its source, it is diverted for drinking, irrigation, and hydro power — putting downstream countries in a vulnerable position. Many countries in water-scarce areas also share lake waters and aquifers. With more people chasing less water, the social, political, and economic impact of water scarcity is becoming a destabilizing force between countries. Even within a country, disputes can break out between political jurisdictions. The mayor of Mexico City, for example, has predicted that conflicts will break out in the Mexican Valley in the foreseeable future if a solution to his city's water crisis is not found soon. And in the United States, a dispute between Nebraska and Kansas over the use of water from the Republican River has been taken all the way to the Supreme Court. Kansas has alleged that Nebraska has allowed unregulated and unrestricted well drilling and pumping in the river basin, which depleted the flow of water into Kansas.

Most border disputes, however, are between countries. In 1997, for instance, Malaysia, which supplies about half of Singapore's water, threatened to cut off that supply after Singapore criticized its government policies. In Africa, relations between Botswana and Namibia have been severely strained by Namibian plans to construct a pipeline to divert water from the shared Okavango River to eastern Namibia. Farther north, Ethiopia plans to divert more water from the Nile — although Egypt depends heavily on that river for irrigation and power. Other tensions have arisen because of Turkey's plan to dam the Euphrates River, which it shares with Syria and Iraq, and Bangladesh suffered greatly when India diverted water from its borders. Bangladesh actually depends on river water that flows from or through India, but in the 1970s, when India was faced with increasing food security problems, it diverted the flow of these rivers into its irrigation systems. Bangladesh was left dry. It took over 20 years for the two countries to sign a water-sharing treaty to end their dispute.

———

In 1992, Slovakia, then a province of Czechoslovakia, ignored objections from environmentalists and started operations on the Gabcikova Dam on the Danube River along the border with Hungary. The Hungarians had been participants in the project, but pulled out in 1989 in response to that country's growing environmental movement. In 1993, the opposing sides agreed to refer the case to the International Court of Justice in The Hague, but much damage had already been done. The water table in the Danube had seriously dropped, drying out thousands of hectares of forest and wetlands and reducing fish catches in the lower Danube by 80 percent.

Back in North America, disputes over the control and use of groundwater beneath the U.S.-Mexican border threaten to create major tensions between the two countries. First, the Hueco Bolson, an aquifer that serves municipal water use from Las Cruces to El Paso to Ciudad Juarez, Mexico, is trending toward depletion. In addition, the U.S. has proposed building a major irrigation canal that would serve California's Imperial Valley. These and many other water-extraction projects threaten to deplete groundwater along the border. Though there is a treaty between the two countries covering surface water, there is unfortunately no agreement covering groundwater, so the disagreements between the countries will have to be settled without benefit of a covenant between the two parties.

At the northern border, conflicts over water use are bound to grow among the 40 million people from eight U.S. states and two Canadian provinces who share the Great Lakes Basin. With water tables falling, the new demands of hundreds of new sprawling communities that lie just outside the basin area (whose demand has outstripped local supply) are straining the Great Lakes system to capacity. William Ruoff, the mayor of Webster, New York (population 2,500), learned just how contentious water issues can be when he placed an ad in the *Wall Street Journal* and the *New York Times*, offering to sell 2 million gallons (about 7.5 million liters) of "crystal clear" well water to the highest bidder. Ruoff backed off when the Great Lakes Governors and Premiers Association informed him that he was offering to sell Lake Ontario water and had no business doing so.

Fears in Canada about U.S. interest in its water are deep and longstanding. In the mid-19th century, the United States first began following a policy of Manifest Destiny, or continental expansion — an obvious threat to Canadian sovereignty. Today, Canadians are more concerned about the inclusion of water in the North American Free Trade Agreement (NAFTA) as a tradable commodity. (See Chapter 7.) Many Canadians believe that American politicians and business leaders view Canadian resources, including water, as continental resources, to be shared as if there were no border. Although Canadians have sovereignty over their land, some fear that if the U.S. runs short of water and Canadians refuse to divert their resources south of the border, Americans might view this as tantamount to a declaration of war. Canadian concerns were not allayed when President George W. Bush remarked in July 2001 — just before the famous G-8 meeting in Genoa, Italy — that he saw Canadian water as an extension of Canada's energy reserves, to be shared with the U.S. by pipeline in the near future.

Tensions that are only potential in North America have already resulted in conflict in the Middle East, where water is perhaps as precious, and as contentious, an issue as anywhere else on earth. Forty percent of Israel's groundwater supply originates in occupied territories and water scarcity has been an issue in past Arab-Israeli wars. In 1965, Syria tried to divert the Jordan River from Israel, provoking Israeli airstrikes that forced Syria to abandon the attempt. Israel diverts water from the Jordan River, leaving the country of Jordan itself with depleted resources. Although armed conflict over water has not arisen between Jordan and Israel, the late King Hussein once said that he would never go to war with Israel *unless* it was over water.

Water scarcity has also increased tensions between Israel and the 2.3 million Palestinians living in the Occupied Territories. Even in recent times of drought, Israel has kept its parks green and grown thirsty crops like cotton by limiting supplies to the 2.3 million Palestinians in the Occupied Territories. While some Israelis refuse to give up their watered lawns and swimming pools, many Palestinians are forced to get their drinking water from tankers, and Israeli per capita consumption of water is three times that of the Palestinians. "They cannot make peace with thirsty people," said Fadel Kaawash, deputy director of the Palestinian Water Authority.

Water can be used as a target in war as well. During the 1991 Gulf War, the United States considered bombing dams in the Euphrates and Tigris rivers north of Baghdad but backed off for fear of high casualties. The Allies also discussed asking Turkey to reduce the Euphrates flow at the Ataturk Dam upstream from Iraq. As it was, they targeted Baghdad's water-supply system while the Iraqis destroyed Kuwait's desalination plants.

In Yugoslavia a 1999 NATO bombing contaminated the massive aquifer that supplies most of eastern Europe with fresh water. Targets included a petrochemical factory making artificial fertilizers, a chlorine-producing factory, a factory for the chemical production of rocket fuel, the municipality of Grocka where a nuclear reactor is situated, and four national parks. Chemicals released into the water table as a result of these bombings will be there for decades, perhaps centuries.

PRIVATE VS. PUBLIC CONTROL OF WATER

Perhaps the most important dispute over fresh water supplies has to do with the increasing role of the private sector in deciding who gets it and why. No sector in the world has become more conscious of the worth of water than the private sector, which sees a profit to be made from scarcity. The result is a fairly new phenomenon: water trading for profit.

Informal, small-scale water trading among farmers is common throughout the nonindustrialized nations of the South and was once frequent in the North as well. These arrangements are made between local farmers and local communities and are based on principles that view water as a common heritage, to be shared on the basis of need. But today, water trading, as carried out by large transnational corporations, is based on principles of profit, which are driving the price of water out of reach of the poor. In addition, when large corporations enter the game, they typically buy up block water rights, deplete water resources in an area, and move on. When Chile privatized water, for instance, mining companies were given nearly all the water rights in that country, free of charge. Today, they control Chile's water market and the shortage of water has served to push up prices.

In California, water rights trading is becoming a very big business. In 1992, the U.S. Congress passed a bill allowing farmers, for the first time in U.S. history, to sell their water rights to cities. Then, in 1997, Interior Secretary Bruce Babbitt announced plans to open a major water market among the users of the Colorado River. The new system would allow interstate sales of Colorado River water between Arizona, Nevada, and California.

Wade Graham of *Harper's Magazine* calls this development "the largest deregulation of a national resource since the *Homestead Act* of 1862" and adds that the only measure that could have topped it would have been the privatization of all U.S. federal lands. Babbitt was counting on the free market to do what politicians and the courts have not been able to do — act as a referee between the many parties laying claims to the Colorado's water. The deals are expected to be small at first, like the arrangement already reached between Nevada and Arizona, with Arizona storing water for Nevada to use in the future. In the long run, the fast-growing areas where the high-tech industry is concentrated will be able to obtain vast quantities of reasonably priced water from what is misconstrued as a virtually limitless source.

A similar experiment in water privatization has already taken place in the Sacramento Valley, and Graham points to it as a warning. For the first time, in the early 1990s, southern California cities and farmers were no longer prevented from buying water directly from farmers in northern California, hoarding it, and selling it on the open market. Large-scale operators helped themselves to huge amounts of water and stored it with the Drought Water Bank until the price was right to sell. A small handful of sellers walked away with huge profits, while other farmers found their wells running dry for the first time in their lives. The results were disastrous: the water table dropped and the land sank in some places.

Graham compares this incident with the Owens Valley tragedy at the turn of the century. The once lush, water-rich Owens Valley was bled dry when water officials from Los Angeles devised a scheme to divert Owens Valley water to southern California. "The Owens Valley scam," writes Graham, "demonstrated that although only a few individuals or corporate entities hold registered water rights, the entire community depends upon those rights . . . Water in California is prosperity, and if the legal right to use it can be privatized and transferred away, then the prosperity of the community may go with it." No portion of the private sector knows this better than the computer industry, which is claiming unfair shares of local water supplies. Computer manufacturers use massive quantities of de-ionized fresh water to produce their goods and are constantly searching for new sources. Increasingly, this search is pitting giant high-tech corporations against economically and socially marginalized peoples in a battle for local water.

Electronics is the world's fastest-growing manufacturing industry according to the Silicon Valley Toxics Coalition. Giants such as IBM, AT&T, Intel, NEC, Fujitsu, Siemans, Philips, Sumitomo, Honeywell, and Samsung have annual net sales exceeding the gross domestic product of many countries. There are currently about 900 semiconductor fabrication facilities (fabs) around the world, where the computer wafers used in computer chips are manufactured. Another 140 plants are now under construction. These plants consume a staggering amount of water. The question is: Where will the water come from? It will have to be derived from the limited amounts available, and that will not happen without conflict. As the

Southwest Network for Economic Justice explains: "In an arena of such limited resources, a struggle ensues between those who have traditionally enjoyed these resources and those newcomers who look at these resources with covetous eyes."

High-tech companies are engaging in mechanisms to capture traditional water rights at low cost, without having to pay for cleaning up contaminated water. These include *water pricing*, whereby industry pressures governments for subsidies and circumvents city utility equipment to directly pump water, thus paying much less than residential water users; *water mining*, whereby companies gain rights to deplete aquifers while driving up the access costs to smaller users such as family farmers; *water ranching*, whereby industry buys up water rights of ranches and farmers; and *waste dumping*, whereby industry contaminates the local water sources and leaves the community with polluted water or a costly cleanup bill.

Despite increasing industrial demand, conservation programs aimed at ordinary people are not applied to industry. The *Albuquerque Tribune* pointed out this irony in a description of a city conservation project: "While some residents tore out their lawns last year (1996) to save water," the newspaper said, "it poured with increasing volume through the spigots of industry." While residents were required to decrease their use by 30 percent, Intel Corporation, a software company based in Albuquerque, was allowed to increase its use by the same amount. In addition, Intel pays only a quarter of what the city's residents pay for their water. Perhaps the most disturbing trend, however, is the deliberate destruction of a local traditional pueblo *acequia* — a collective system of agricultural water distribution — to feed the voracious appetite of the high-tech giants.

Under the new commercial system, water is separated from the land it belongs to and transported great distances. This is anathema to the local Indigenous ways and makes no long-term economic or ecological sense. Says John Carangelo, a mayordomo of the La Joya Acequia Association, "In New Mexico, where the total finite supply of water is allegedly fully appropriated, the location of a high-tech industry is dependent on the purchase of existing water rights. This high demand for water and their vast financial resources makes water a valuable commercial product." He warns that water trading could hollow out rural America.

~~~

As the planet dries up and water supplies are bought up by private interests, we have begun moving into a new economic configuration, where sprawling cities and agribusiness operations thrive and the wells of private citizens and local farmers run dry. Old ways of wasting water — like the rights trading that benefited a few but devastated the Owens Valley in southern California — are being revived, though they were demonstrated failures in the past. Meanwhile, in Third World countries, where children are already dying of thirst, the World Bank and the International Monetary Fund make privatization of water services a condition of debt rescheduling, and the poor soon find they are unable to pay for the skyrocketing costs of water and sanitation services. What lies ahead is a world where resources are not conserved, but hoarded, to raise prices and enhance corporate profits and where military conflicts could arise over water scarcity in places like the Mexican Valley and the Middle East. It's a world where everything will be for sale.

PART II

THE POLITICS

EVERYTHING FOR SALE

*How economic globalization
is driving the world water crisis*

I f water is essential to life itself, then is it simply a basic human need or is it a fundamental human right? That was the subject of the debate that burst out on the floor at the World Water Forum where 5,700 people gathered together for four days in The Hague in March 2000. The title of the conference sounded like an official United Nations meeting about conserving world water resources, but it wasn't. The World Water Forum was anything but. It was convened by big business lobby organizations like the Global Water Partnership, the World Bank, and the leading for-profit water corporations on the planet, and the discussions focused on how companies could benefit from selling water to markets around the world.

It's true that UN officials were in attendance, along with a Ministerial Conference attached to the event, where over 140 governments were represented. But they were not in charge. The main players were some of the world's largest corporations — the self-proclaimed saviors of the global water crisis. They included not only global water giants like Vivendi and

Suez, but also big food-processing conglomerates like Nestlé and Unilever, the purveyors of bottled water.

The debate over whether water should be designated a "need" or a "right" was not simply a semantic one. It went to the heart of the question of who should be responsible for ensuring that people have access to water — the essence of life itself. Would it be the market or the state, corporations or governments? The debate would not, in all likelihood, have taken place had it not been for the presence of a small group of civil society groups. Working together under a common banner that became known as the "Blue Planet Project," representatives of environmental, labor, and public interest groups from both industrialized and nonindustrialized nations insisted that water be recognized as a universal human right.

But the conveners of the World Water Forum had a different agenda. They wanted water to be officially designated as a "need" so that the private sector, through the market, would have the right and responsibility to provide this vital resource on a *for-profit* basis. If, on the other hand, water was officially recognized as a universal human right, then governments would be responsible for ensuring that all people would have *equal access on a nonprofit basis*. In the end, the government representatives deferred to the corporate interests of the forum's sponsors. A statement, signed by government officials attending the Ministerial Conference, declared that water was a basic "need." It said nothing about water being a universal "right."

⌒

The story of what happened at the World Water Forum is the story of the separation of water from the land and from "the commons" to which it belongs. It is also a denial of historic benchmarks for democracy enshrined halfway through the 20th century. After all, at that time, the Universal Declaration of Human Rights, along with its accompanying International Covenants on Economic, Social and Cultural Rights and on Civil and Political Rights were adopted as the cornerstones of the United Nations. They were among the crowning achievements in the struggle for democracy that had characterized much of the past two hundred years. Yet at the dawn of the 21st century, something as fundamental as water is no longer recognized as a universal right by the dominant economic and political elites. Being designated a need, water has been subjected to the supply and demand forces of the global marketplace, where the distribution of resources is determined on the basis of the ability to pay.

To fully understand these dynamics, we need to look at the forces of economic globalization that are reshaping the lives of peoples, communities, and nations today. After all, the world in which the water crisis is exploding lives under the sway of a global economy run by transnational corporations. In this age of economic globalization, governments have largely abandoned their responsibilities to act in the public interest or the common good, and increasingly, the rights of corporations supersede those of citizens. We need to come to grips with these dynamic forces of globalization, in order to understand the causes of the impending world-wide water crisis. Only then will we be able to find solutions.

ECONOMIC GLOBALIZATION

The dominant development model of our time is economic globalization, a system fueled by the belief that a single global economy with universal rules set by corporations and financial markets is inevitable. Economic freedom, not democracy or ecological stewardship, is the defining metaphor of the post–Cold War period for those in power. As a result, the world is going through a transformation as great as any in history. At the heart of this transformation is an all-out assault on virtually every sphere of life. In this global market economy, everything is now up for sale, even areas of life once considered sacred, such as health and education, culture and heritage, genetic codes and seeds, and natural resources, including air and water.

The roots of economic globalization go back over five hundred years to a time when the empires of Europe competed with each other in their race to seize control over valuable resources like the gold, silver, copper, and timber that nature had stored in Asia, Africa, and the Americas. Back then, huge shipping enterprises like the Hudson's Bay Company and the East India Company were among the first examples of what we call transnational corporations today. Licensed to operate through royal charters, these original transnational enterprises were mandated to scour large areas of the world in search of resources to fuel the profitability of their commercial empires. While the resources targeted for economic development have changed in response to new technologies over the centuries, the basic model of economic globalization has remained much the same.

In our times, this model of economic globalization has been accelerating at a rapid pace, particularly since the fall of the Berlin Wall. Before this, and for most of the 20th century, the global economy was divided between two competing models: communism and capitalism. Symbolically, at least, the collapse of the Berlin Wall and the end of the Cold War marked the triumph of capitalism over communism and the termination of this bipolar economy. From this point onwards, capitalism has reigned supreme in the global economy. As the dominant institutions of global capitalism, transnational corporations went on a rampage, opening up markets and spreading their operations into the four corners of the planet.

After World War II, the United States emerged as an industrial superpower. It was producing so many consumer goods that it wanted to open up new global markets and promote free market systems and values around the world. This ideology took root in the following decades and came to be known as the Washington Consensus, a term coined in 1990 by John Williamson of the Institute for International Economics, a Washington-based conservative think tank. The so-called consensus featured massive government deregulation of trade, investment, and finance and became the official ideology of the new world order. According to this doctrine, it is essential that capital, goods, and services be allowed to flow freely across borders around the world, unfettered by government intervention or regulation. At the core of this ideology is the belief that the interests of capital take priority over the rights of citizens. For these reasons, the Washington Consensus has been called "democracy delayed," since it largely rejects the precedence of peoples' democratic rights, the philosophy that lies at the heart of the Universal Declaration of Human Rights and its accompanying covenants.

This doctrine of economic liberalization itself is based on principles advanced by the Trilateral Commission. Initially formed in the early 1970s, the Trilateral Commission was designed to bring together 325 of the world's top economic and political elites — including the CEOs of some of the largest corporations and banks, presidents and prime ministers of the leading industrialized countries, senior government officials, like-minded academics, and public opinion makers in the media. In one of their first major reports, entitled "The Crisis of Democracy," the Trilateralists declared that the central political problem of our times had to do with the current model of governance, claiming that there was an "excess of democracy" in the system.

The Trilateralists went on to develop their own blueprint for restructuring the global economy and its governing institutions: the International Monetary Fund (IMF) and the World Bank, established in 1944 by the Bretton Woods Conference following World War II, and the General Agreement on Tariffs and Trade (GATT), formed in 1947, which was replaced by the World Trade Organization (WTO) in 1995. To create a borderless world, the Trilateralists repeatedly called for massive reductions in tariff and non-tariff barriers to global trade, especially in textiles, clothing, footwear, electronics, steel, ships, and chemicals. In response to the growing debt burden of the nonindustrialized countries of the South, they proposed that the IMF and the World Bank impose "Structural Adjustment Programs" (SAPs) on these nations, requiring them to radically change their economic and social policies in accordance with global free market priorities.

By promoting this agenda for restructuring the global economy, the Trilateralists were able to rapidly accelerate the process of economic globalization, especially in the last decade of the millennium. In so doing, they both ignored and outflanked the United Nations. Instead, the Trilateralists saw themselves as self-proclaimed leaders with a mission to create an ideological consensus for building a new world order. As a consequence, a new global royalty now centrally plans the world economy, creating human hardship and destroying nature as it proceeds.

Transnational Corporations

Two decades ago, the UN reported that there were some seven thousand transnational corporations in the world, but today the numbers extend to well over forty-five thousand. The top two hundred, says the Washington-based Institute for Policy Studies, are so large and powerful that their combined annual sales are greater than the sum total of the economies of 182 of the 191 countries in the world. What's more, they have almost twice the economic clout in terms of annual income as the poorest four-fifths of humanity on the planet. Of the largest 100 economies in the world today, 53 are transnational corporations rather than nation-states.

Based on the Global Fortune 500 ratings for the year 2000, ExxonMobil, for instance — currently the largest transnational conglomerate in the world — has more total revenue than all but 22 nation-states on the planet, and Wal-Mart, ranked second, is larger than the economies of 178 countries. General Motors' economy is bigger than that of either Hong Kong or Denmark, and Ford Motor Company has annual sales greater than the national revenues of either Norway or Thailand. Royal Dutch/Shell's yearly revenues outstrip those of Poland and South Africa, and British Petroleum has annual revenues larger than any of Saudi Arabia, Finland, and Portugal. Also on the mega-conglomerate list are Mitsubishi, Toyota Motors, and Mitsui, each of which has annual sales greater than the revenues of countries like Israel, Egypt, and Ireland.

At the same time, the annual sales and profits of the top two hundred global corporations have dramatically outpaced world economic growth. According to the Institute for Policy Studies, the total sales of the world's top 200 corporations grew at a pace of 160 percent, while their combined profits jumped by as much as 224 percent between 1983 and 1997. During the same period, total world economic growth increased by only about half that amount — at 144 percent. In addition, there has been a tremendous boom in mergers and the concentration of corporate wealth. In 1998 alone, merger deals worth US$1.6 trillion were made, a 78 percent increase over 1997.

Through these mergers, global production and marketing is becoming concentrated in the hands of fewer and fewer transnational empires. ExxonMobil and British Petroleum-Amoco, for example, now control most of the world's petroleum and refining. Four U.S. corporations (International Paper, Georgia-Pacific, Kimberly-Clark, and Weyerhaeuser) dominate worldwide forest and paper production. Meanwhile, giant retailers like Wal-Mart have led the way in creating global shopping malls through superstore chains designed to sell the largest range of retail goods. And two

European conglomerates, Nestlé and Unilever, control the lion's share of the global food processing market, while other brand name food companies like General Foods, Kraft, Pillsbury, Philip Morris, Del Monte, and Procter & Gamble have merged their own operations and expanded their marketing strategies on a global basis.

Yet the most provocative innovations have probably taken place in the service industry, where for-profit corporations have been taking control of public services like health care, education, and water, previously delivered by governments and their agencies. Although the major pharmaceutical corporations have cornered a piece of the health care market, one of the most significant players is an enterprise resulting from a merger between two large U.S. hospital chains in the U.S.: Columbia and Health Trust. The result was the world's largest for-profit health care corporation with annual sales exceeding those of Eastman Kodak or American Express. Meanwhile, in the field of public education, the formation of the New American Schools Development Corporation, designed to funnel corporate finances into profit-oriented elementary schools in the United States, has been spearheaded by transnationals like AT&T, Ford, Eastman Kodak, Pfizer, General Electric, and Heinz.

And now water services are being targeted by for-profit enterprises. Two French-based transnational conglomerates, Vivendi and Suez, are now referred to as the General Motors and Ford Motor Company of the world water industry. In 2000, Vivendi and Suez were ranked 91st and 118th, respectively, in the Global Fortune 500. Between them, they own, or have controlling interests in, water companies in over 130 countries on all five continents, and combined, they currently distribute water services to more than a hundred million people around the world.

Yet these corporations, their managers, and their investors are made legally immune by "acts of incorporation" from any corporate harms done to societies, people, or the environment. What is legally classified as the "publicly traded, limited liability corporation" is virtually guaranteed immunity under a vast body of corporate law that has been built up nationally and internationally over the course of more than a century. As Canadian philosopher and activist John McMurtry puts it, "This is the legal armor around the global corporate system which gives them unaccountable impunity for whatever damage or crime they impose on individuals, societies or environments around the world."

COMMODIFYING NATURE

One of the prime driving forces behind transnational corporations and the expansion of the global economy has been the "growth imperative," and in recent years, people have begun to recognize that this principle is on a collision course with Nature itself. In their classic work *For the Common Good*, Herman E. Daly and John B. Cobb demonstrated that orthodox economics, based on the growth imperative, was rooted in a narrow definition of "capital" as human-made assets like goods and services, machines and buildings. Left out of the equation, said Daly and Cobb, was what they called "natural capital," the resources of the earth which make all economic activity possible. But the carrying capacity of the planet's ecosystem has its limits, especially given the rapid pace at which the natural world is being destroyed by industrial agriculture, deforestation, desertification, and urbanization. At this rate, Daly and Cobb warned, the collision could come within the next generation.

In India, feminist, physicist, and ecological activist Vandana Shiva takes the concept one step further by arguing that the growth imperative amounts to "a form of theft" from Nature and people. It is true, says Shiva, that cutting down natural forests and converting them into monocultures of pine for industrial raw materials generates revenues and growth. However, it also robs the forest of its diversity and its capacity to conserve soil and water. And by destroying the forest's diversity, communities of people who depend on the original forest for their sources of food, fodder, fuel, fiber, and medicine, as well as for protection from drought and famine, are also being robbed. Shiva goes on to show how the growth imperative, applied to industrial agriculture, also fails to provide more food, reduce hunger, or save natural resources. Industrial agriculture also operates as a form of theft from Nature and the poor. She would argue that the same can be said of the construction of huge, high-tech power dams and the diversion of river systems.

At the heart of this critique lie concerns about the increasing commodification of Nature and of life itself. There was a time not so long ago when certain aspects of life and Nature were not considered commodities to be bought and sold in the marketplace. Some things were not for sale — things like natural resources (including air and water), genetic codes and seeds, health, education, culture, and heritage. These, and other essential elements of life and Nature, were part of a shared inheritance or rights that belonged to all people. In other words, they belonged to "the commons." In India, for example, space, air, energy, and water have traditionally been viewed as "incapable of being bound into property relations." They were not to be treated as private property but as "common resource property," and they were not to be subjected to market forces such as the laws of demand and supply. On the contrary, these dimensions of common life were deemed to be of universal importance, and in many respects, they were considered sacred. As such, they were to be protected and preserved by governments through the public sector or more directly by local communities themselves.

The commodification of water, in particular, is seen to be a direct assault on the commons. In a report by the Research Foundation for Science, Technology and Ecology (a New Delhi-based NGO directed by Dr. Vandana Shiva), water in India is understood to be "life itself, on which our land, our food, our livelihood, our tradition and our culture depends." As "the lifeline of society," water is "a sacred common heritage . . . to be worshiped, preserved and shared collectively, sustainably used and equitably distributed in our culture." In the traditional teachings of Islam, for example, the "Sharia," or "way," originally meant the "path to water," and the ultimate basis for the "rights of thirst," which apply to both people and Nature. Based on these traditional spiritual and cultural traditions, says the foundation, local communities in India have developed "creative mechanisms of water management and ownership through collective and consensual decision making processes" designed to ensure "sustainable resource use and equitable distribution."

But now, in this age of economic globalization, water is being commodified and commercialized in India, says the foundation, with alarming results. Under pressure from the IMF and the World Bank to secure revenues to pay down their debt load, the Government of India has been selling water rights to global water corporations, including Suez and Vivendi, and to major industries requiring heavy water use for their production operations. As a result, Indigenous local traditions of water management and harvesting are bypassed, giving way to increased "commercialization and over-exploitation of scarce water resources." What we are witnessing, reports the foundation, is "the enclosure of a hitherto 'common property resource' into private commodities." The impacts of this commodification of water amount to "irreversible losses to our environment and livelihoods of our people." What's more, these trends are being experienced not only in India but throughout most of the Third World.

Indeed, the commodification, not only of water, but of other parts of Nature and of life itself, is a distinguishing feature of corporate-led globalization today. What was once understood to be "the commons" has become the last frontier in the expansion of global capitalism. As transnational corporations conquer markets around the world, new industries are emerging to commercialize the remaining elements of our common life. A prominent example in recent years has been the biotechnology industry. Presenting themselves as "life science" industries, major biotech corporations like Monsanto and Novartis have been turning seeds and genes into commodities to be bought and sold as genetically engineered food and health products in global markets. Similarly, the global water giants have been busy transforming this life-giving resource into a commodity to be sold on a for-profit basis to those who have the ability to pay. In short, everything is now for sale to the highest bidder, including seeds, genes, and water. And the fundamental contradiction underlying the commodification of water was articulated by no less than the CEO of the global water giant Suez. "Water is an efficient product," said Gérard Mestrallet. "It is a product which normally would be free, and our job is to sell it. But it is a product which is absolutely necessary for life."

PRIVATIZATION SCHEMES

As one of the cardinal features of the Washington Consensus, the private takeover of public institutions and enterprises has become the prime instrument for the commodification of water. Public services like the delivery of water, traditionally by municipal governments in most countries, are taken over by corporations, often foreign owned, in the interests of making a profit. Through this privatization process, water is turned into a commodity, priced, put on the market, and sold, usually on the basis of ability to pay.

Water privatization generally occurs in one of three forms. First, there is the complete sell-off by governments of public water delivery and treatment systems to corporations, as has happened in the United Kingdom. Second, there is the model developed in France, whereby water corporations are granted concessions or leases by governments to take over the delivery of the service and carry the cost of operating and maintaining the system, while collecting all the revenues for the water service and keeping the surplus as a profit. Third, there is a more restricted model, in which a corporation is contracted by the government to manage water services for an administrative fee, but is not able to take over the collection of revenues, let alone reap profits from surpluses. While all three forms contain the seeds of privatization, the most common one is the second model, often referred to as "public-private partnerships."

Shifting from public to private systems introduces, of course, a completely different set of commercial imperatives into water-service delivery. Although the water industry insists on "full cost recovery" in taking over a concession, this usually includes profit margins. After all, the owners and shareholders of the privatizing corporation are driven by demands for profits and dividends, which in turn, are generally redistributed for investment globally in other divisions of the corporation's overall operations. Maximizing profit is the prime goal, not ensuring sustainability or equal access to water. Management of water resources, therefore, is based on market dynamics of increasing consumption and profit maximization, rather than on long-term sustainability of a scarce resource for future generations. As a result, the price that a corporation is prepared to pay for a water concession depends on the revenue- and profit-generating stream that can be expected from the deal.

Ensuring such profit-generating revenues for the corporation means charging higher prices for water services. Since water services were privatized in France, for example, customer fees have increased by 150 percent.

In England, according to Public Services International (PSI), the France-based organization representing public sector unions around the world, customer water charges jumped 106 percent between 1989 and 1995 while profit margins for the private water companies increased 692 percent. Likely as a result of these price hikes, the number of British customers who have had their water disconnected has risen by 50 percent. What's more, the price of water service charges by private corporations or by "public-private partnerships" (PPPs) is consistently higher than those managed by municipal governments. The Public Services International Research Unit (PSIRU) (a group separate from but related to PSI) has carried out studies showing that price charges for privatized or PPP-delivered water services in France was 13 percent higher than those of municipal governments in 1999 alone. And in nonindustrialized countries, the price impacts of privatization are much more severe. In India, for instance, some households are compelled to pay a staggering 25 percent of their incomes on water.

Yet cash-starved governments are rapidly turning to water privatization as a solution to their own financial problems. Due to substantial cuts in corporate taxes almost everywhere, for example, many local governments no longer have the tax revenues necessary to cover their operations, let alone public services. As a result, governments and public institutions find themselves plagued by crippling debt and deficit problems. To make matters worse, deteriorating water infrastructure like leaky pipes has become a major problem in both industrialized and nonindustrialized countries — especially in inner cities where government spending on public works has been drastically curtailed. In Boston, Massachusetts, for instance, 40 percent of municipal water was being lost through broken pipes until recently, but the cost of rebuilding this infrastructure was astronomical. And in countries of the South, over 50 percent of municipal water and 60 to 75 percent of irrigation water is lost because of leaky pipes and related problems. At first glance, privatizing local water systems may make sense to cash-strapped governments. The proceeds of these sales can help alleviate their financial debt, and in the process, they do not have to take responsibility for making improvements in water infrastructure.

However, these privatization schemes are generally financed through governments and public institutions. According to a World Bank report, this kind of financial support includes "cash contributions during the construction period; subsidies during the operating period, e.g. in the form of non-refundable grants; and a favorable tax regime — including tax holidays, refunding of tax on construction and operating costs." In order to minimize the risk transferred to the private sector, PSIRU reports that public authorities are expected to provide financial guarantees. These include both loan guarantees and profit guaranteees — that is, development banks often require government guarantees before money is lent to privatized operations like water service delivery, and many water concession contracts include clauses requiring governments to guarantee that the private operators will receive profits during the contract period. Profit guarantees, for example, were built into the water concession contracts for Cochabamba in Bolivia, Plzen in the Czech Republic, and Szeged in Hungary. These government financial guarantees, of course, come out of the pockets of taxpayers.

Once privatization schemes are implemented, public controls diminish substantially — even though the public has paid for financial guarantees. Most privatized water systems involve long-term concession contracts lasting between 20 and 30 years, and these contracts are extremely difficult to cancel, even if unsatisfactory performance can be demonstrated. In cases where public authorities have tried to cancel contracts (in places like Valencia, Spain; Tucumán, Argentina; Szeged, Hungary; and Cochabamba, Bolivia), the global water corporations have either threatened or carried out threats to sue for damages, thereby making the cancellation incredibly expensive. In the case of Cochabamba, the Bechtel Corporation is suing the Bolivian government for nearly US$40 million at the World Bank's International Centre for Settlement of Investment Disputes. Claiming "expropriation rights" under a Bilateral Investment Treaty between Bolivia and the Netherlands, Bechtel is using its Dutch-based holding company to gain the right to sue Latin America's poorest country directly. To prove that Bolivia is ready to be a "good player" in the game of economic globalization, there is considerable pressure on the government to settle the dispute out of court by paying Bechtel the compensation it is demanding.

The water service industry has also been plagued by a lack of public transparency. At the spring 2000 World Water Forum in The Hague, for instance, one corporate executive claimed that "as long as water was coming out of the tap, the public had no right to any information as to how it got there." And in Canada, the residents of Walkerton, Ontario, were stunned when they heard, after seven people had died from *E. coli*-contaminated drinking water in their town, that the private testing lab, A&L Laboratories of Tennessee, was no longer supposed to report the contamination to provincial government authorities. In 1999, Ontario's Harris government had dropped testing for *E. coli* from its Drinking Water Surveillance Program, and a year later, it had closed the program down entirely. By not reporting to provincial authorities, A&L Laboratories was complying with the new legal requirements mandated by the Harris government. As a result, a lab spokesman said that the test results were "confidential intellectual property," and, as such, belonged only to the "client" — namely, the public officials of Walkerton. In other words, privatization had inevitably led to less public accountability because less reporting by companies to government authorities was required.

FINANCIAL SPECULATION

Although transnational corporations are the prime institutions of economic globalization, it is mainly financial speculation on capital markets that fuels the expansion of the global economy. The more that water becomes a profitable commodity to be bought and sold in global markets, the more it becomes the target of foreign speculators in financial markets. And given the increasing scarcity of available fresh water supplies, the price of water could skyrocket as a result of investors speculating on commodity markets.

Indeed, the global economy today is fueled largely by a financial casino, in which most investors have become speculators and gamblers. Instead of buying long-term shares in companies producing goods and services, investors now tend to put their money into mutual funds where they can speculate or gamble on fluctuations in commodity prices or the value of currencies. *Speculative investment*, in other words, has supplanted *productive investment* as the engine of the global economy. An average of close to US$2 trillion dollars a day is moved around the world through this global casino, primarily in the form of speculative investments. With one keystroke, commodity and currency traders can move vast sums of money all over the world, using digital electronic information systems to track price fluctuations. Temporarily parking their money in markets that offer high-end, short-term returns, commodity speculators can suddenly withdraw their money and move it elsewhere, thereby destabilizing a country's economy.

Says David Korten, writer and former senior advisor to USAID (the U.S. Agency for International Development, which provides "development assistance" to other countries): "The world is now ruled by a global financial casino staffed by faceless bankers and hedgefund speculators who operate with a herd mentality in the shadowy world of global finance. Each day, they move more than two trillion dollars around the world in search of quick profits and safe havens, sending exchange rates and stock markets into wild gyrations wholly unrelated to any underlying economic reality. With abandon they make and break national economies, buy and sell corporations and hold politicians hostage to their interests."

As for other commodities, the ground is now being laid for a commodity futures market involving speculation on water prices. At a March 1998 conference in Paris, the UN Commission on Sustainable Development proposed that governments turn to "large multinational companies" for capital and expertise and called for an "open market" in water rights and an enlarged role for the private sector. The UN promised to mobilize private funds for the vast investments needed to set up networks and treatment plants and to pay for the technology required to ensure future water supplies. Meanwhile, investment speculators and corporations have been buying up block water rights in farmland areas to sell in bulk form to thirsty cities, especially in the United States. In 1993, for instance, the billionaire Bass brothers of Texas quietly bought up forty thousand acres (about sixteen thousand hectares) of Imperial Valley farmland in order to sell water to the city of San Diego, California, although the project later fell through.

Recently, the term "water hunters" has been used to describe this new breed of entrepreneurs. From the rainforests of the Amazon to the aquifers beneath the desert regions of Africa, these hunters mine the planet for fresh water sources to put on sale in the boutique markets of Paris and New

York. As blue gold becomes more scarce, we are likely to hear a great deal more about these water hunters.

In January 1999, U.S. Filter Corp., now a subsidiary of the global water giant Vivendi, bought a ranch and fourteen thousand acre-feet of water north of Reno, Nevada, which it intends to divert by pipeline to Reno for commercial sale. The local community of Lassen County says it will be left without its lifeblood. In early 2001, the Metropolitan Water District of Los Angeles contracted to buy as much as 47 trillion US gallons (about 178 trillion liters) of water from the state's largest farming company, Cadiz Inc., which plans to have it pumped from an aquifer deep under the Mojave Desert. Yet environmentalist Tony Coelho, formerly a powerful Democratic congressman and a chairman of Al Gore's year 2000 presidential campaign, maintains that this water source is so valuable that no dollar figure can be put on it. Keith Brackpool, the British entrepreneur who runs Cadiz, differs. "If you do the math," he says, "the price of our water just soars." In other words, Brackpool can buy the water cheaply and sell it at high rates to thirsty L.A. industries and residents.

It's not surprising, then, that California's Governor Gray Davis has said: "Water is more precious than gold." Clearly, buying and selling block water rights have become big business in California. In a private market, however, the superior purchasing power of large cities such as Los Angeles and of corporations such as Intel could force the cost of water up far enough to make water so expensive that it's out of the reach of small farmers, towns, and Indigenous peoples.

At the same time, capital is being pooled to finance massive pipeline schemes for the delivery of water and energy around the world. According to the *Guardian Weekly*, General Electric has joined forces with the World Bank and international investor George Soros to invest billions of dollars in a "Global Power Fund" that would be used to finance major water and energy schemes. This is the same George Soros who, in 1992, bet Britain's then prime minister, John Major, that financiers were more powerful than political leaders. To win his bet, Soros sold $10 billion worth of British pounds in global finance markets for a $1 billion profit, thereby single-handedly forcing a devaluation of the British pound and dismantling a new exchange rate system that had been proposed by the European Union.

INTERNATIONAL COMPETITIVENESS

All these efforts to prepare the way for more and bigger circles of global trade are the result of a philosophy put in motion by the Washington Consensus, the post-World War II ideology that aimed to create one single, unified global economy, based on the doctrine of international competitiveness. According to this guiding principle, the name of the game is primarily to produce goods and services for export markets rather than the domestic market and local development needs. To be internationally competitive, national governments are compelled to eliminate all barriers to the free flow of capital, goods, and services, including environmental regulations designed to protect natural resources like water.

Driven by this doctrine, the growth of global investment and trade over the past three decades has been staggering. Between 1970 and 1992, according to the *World Investment Report*, foreign direct investment by transnational corporations in nonindustrialized countries grew twelvefold. In the next five years (1992–1997) it tripled again, rising to US$149 billion out of a worldwide total of US$400 billion in foreign direct investment. The corresponding push to open markets worldwide by promoting a combination of foreign imports and export production has generated a similar explosion in the volume of world trade. Reports by *World Economic Outlook* show that worldwide trade has grown from US$380 billion in 1950 to US$5.86 trillion in 1997 — a fifteenfold increase in less than half a century.

The imperative of export production has left an increasingly large and more damaging ecological footprint on the planet. To produce for global markets, big fish-trawling vessels have virtually depleted the stocks in many commercial fishing grounds. Meanwhile, the massive logging operations of the timber giants now threaten more than 70 percent of the world's largest virgin forests, and the mining industry currently strips more of the earth's surface annually than rivers do through natural erosion. And as noted in Chapter 2, production of cash crops for export has also resulted in severe ecological damage, including soil erosion, water depletion, and chemical contamination. The drive to export has greatly intensified the exploitation of natural resources in the South — 100 percent of Botswana's diamond mining is for export; 99 percent of Burundi's coffee; 93 percent of Costa Rica's bananas; 83 percent of Burkina Faso's cotton; 71 percent of Malawi's tobacco; 50 percent of Malaysia's timber; and 50 percent of Iceland's fish catch.

To gain a competitive advantage in global markets, industrialized and nonindustrialized countries alike feel compelled to dismantle environmental regulations, including water safeguards. Responsible management of the environment by governments through laws and regulations is frequently viewed as a liability that decreases international competitiveness. Laws restricting the bulk exports of water or the privatization of water services or the construction of hydro power dams on certain rivers are frequently labeled by transnational corporations as "unfair barriers" to international trade and investment. In a globally competitive economic climate, transnational corporations will threaten to withdraw their investment plans in a given country unless the government changes the environmental regulation in question. As a result, many environmental regulations have either been overturned or left unenforced while new ecological safeguards have been prevented from seeing the light of day.

Meanwhile, the global climate for economic competitiveness has intensified demands to turn water into a tradable commodity. For those willing and able to pay top dollar for water on an emergency basis, there are currently barges carrying loads of fresh water to islands in the Bahamas, while tankers deliver water to Japan, Taiwan, and Korea. If plans to establish a European Water Network are realized, alpine water could be flowing into Spain or Greece, rather than Vienna's reservoirs, within a decade. And the trade in bottled water, now estimated to be worth US$22 billion annually, has become one of the fastest-growing and least-regulated industries in the world. Since 1995, bottled water sales have skyrocketed, at a growth rate of over 20 percent annually. In 2000, close to 89 billion liters (about 23.5 billion US gallons) of water were bottled and traded globally.

The explosion of worldwide trade has also generated mass transportation technologies that are causing damage to waterways. Corporate plans are being put together for the mass export of bulk water by diversion, by pipelines, and by supertankers. Increased global shipping not only multiplies the amount of waste dumped directly into lakes and oceans, but the dredging operations required for the construction of ports and canals destroy coastal habitat as well. What's more, the trend lines indicate acceleration. Worldwide shipping, which currently accounts for 90 percent of all trade in goods, is expected to increase by 85 percent between 1997 and 2010, and major ports like Los Angeles predict a doubling of cargo business over the next 25 years. In another ecological intrusion in the making, there are plans to open up the interior of Latin America to world trade by constructing a mammoth new water system that will channel 3,400 kilometers (about 2,100 miles) of the Paraguay and Paraná rivers for shipping operations. While the project is presently on hold, environmentalists fear it could be reactivated at any time. Meanwhile, China has signaled its intent to become a big player in global trade by initiating work on a huge US$1 billion project to divert water from the Yangtze River to Beijing through tunnels, draining water from cities and towns en route.

What's more, world trade rules have been designed in such a way as to protect the rights of global water corporations, the privatization of water services, and bulk exports of fresh water resources. International trade regimes like the North American Free Trade Agreement (NAFTA) and the World Trade Organization (WTO) have already declared water to be a tradable commodity by classifying it as a commercial "good," a "service," and an "investment." What this means, in effect, is that if a government were to put a ban on the sale and export of bulk water or to prevent a foreign-based water corporation from bidding on a concession for the private delivery of water services, it could be challenged as being in violation of international trade rules under the WTO or NAFTA. Both these trade regimes, in turn, contain enforcement mechanisms designed to ensure that their rulings on trade disputes are binding on their member governments. (See Chapter 7.)

CORPORATE STATES

All over the world, governments have failed to take bold measures to protect the commons and offset the impending water crisis. While some tentative steps have been taken to recognize the severe water shortages looming on the horizon, governments have so far failed to develop a comprehensive analysis of the situation, let alone find solutions that protect the fundamental rights of people and nature to water.

Everything for Sale

To be sure, there have been some noteworthy success stories in reclaiming rivers, lakes, and estuaries choked with sewage and industrial pollution. In the United States, the Hudson River, once given up for dead, now abounds with life. Through joint action by governments in both Canada and the U.S., concerted measures have been taken to partially recover the Great Lakes by curbing the dumping of phosphorus and municipal sewage into the lakes. There is also evidence that conservation efforts in Europe and North America have been effective in reducing household and industrial water use, thereby helping to slow the rate at which water is being withdrawn from aquifers and other sources. According to the U.S. Geological Survey, water use has actually dropped in some regions and industrial sectors in the U.S. by 10 to 20 percent since 1980.

Apart from these advances, however, the picture has been bleak. According to the United Nations, governments in both industrialized and nonindustrialized countries give low priority to water issues and institutions, and funding for research and solutions is abysmally inadequate. Cash-strapped governments are faced with deteriorating water infrastructure, including broken and leaky pipes, but they lack the billions of dollars needed to do repairs. The Canadian government, for example, now estimates it will cost US$53 billion to upgrade Canada's deteriorating water infrastructure alone. In addition, governments are giving up their right and abdicating their responsibility to protect their water heritage by default. Most governments have very few laws and regulations on the books governing their water systems, let alone policies and programs to deal with the growing pressures to privatize, commercialize, and trade water.

What's more, governments are directly culpable themselves in that public funds are used to subsidize corporations and industries that exacerbate the water crisis. In the United States and elsewhere, governments subsidize the water-guzzling high-tech industry. For example, the city of Austin, Texas, not only awards tax breaks to high-tech companies (including a recent US$125 million break for Samsung), but also reduces the water rates it charges industry to less than two-thirds of what residents pay. In New Mexico, Intel recently received a tax subsidy of US$8 billion through an industrial revenue bond, plus an additional US$250 million in tax credits and other subsidies. Elsewhere around the world, governments

98

continue to massively subsidize the global transportation system that facilitates expanding world trade and economic globalization. If the full cost of shipping consumer goods across oceans for assembly and then back again to markets was reflected in the final price, the volume of world trade would decline significantly.

Yet none of this should come as a complete surprise. After all, democratic governance itself has been completely undermined by the increasing political power of transnational corporations. Ever since the Trilateralists declared that modern systems of governance today suffer from "an excess of democracy," corporations have been devising more effective strategies and mechanisms for using their immense economic clout to wield political power and influence the governments of nation-states around the world. Armed with their own policy think tanks and legal and public relations firms, corporations are well equipped to develop their own policy and legislative agendas on major public issues. Fortified by sophisticated lobbying machinery, major corporate players work together through big business coalitions to promote their own laws, policies, and programs. In effect, the substantial donations that corporations regularly make to parties and campaigns often serve as an "insurance policy" to try to make sure that their political agendas will be implemented by the government once it is elected.

Over the past quarter-century, transnational corporations have successfully managed to reinvent government in their own image. The models of governance that characterized much of the postwar period in the 20th century, including the social welfare state and the national security state, have been replaced by a new model — namely, the corporate security state. In this age of economic globalization, the primary role of the state is to provide a secure place and climate for profitable transnational investment and competition. In other words, "investor security" is considered to be the prime organizing principle of governments: the priority of governance is to provide security for corporations, not citizens. And if the property and investments of corporations should be seriously threatened by workers or communities, the state would be obliged to invoke police action to defend and protect the rights of investors with armed force.

Indeed, the collapse of the Cold War did not mean the end of global insecurity. Instead, a new battle has now been woven into the processes of economic globalization itself, where corporate-driven trade, finance, and investment are being promoted by those who will benefit from them, to the detriment of others. As Ursula Franklin, Canadian scholar, environmentalist, and long-time peace activist puts it: What we have now is an "economic war" in which the new "enemy" is people and Nature, and the new territories of occupation are "the commons" (those not-for-profit spaces we "hold in common" in a democratic society). Says Franklin, we are living under a military-style occupation with "puppet governments" running the country on behalf of the corporations and their "armies of marketeers." This is the corporate security state that now shapes the political life of nations and peoples in an era of global capitalism. As struggles for control of scarce water resources continue and even intensify, we can expect to see much more of the corporate security state in action. But meanwhile, the rush to profit from thirsty citizens continues, and transnational corporations from many parts of the world are racing to get in on the bonanza.

GLOBAL WATER LORDS

*How transnational water
corporations are commodifying
the earth's water for profit*

On a cool, sunny morning in Johannesburg, South Africa, Dr. David McDonald, Director of the Municipal Services Project (MSP), calmly shared the results of an impact study he had just completed in May 2001 on the privatization of water and sanitation services in Buenos Aires, Argentina. To date, this had been the largest private concession in the world, involving the two largest global water corporations, Suez and Vivendi, with Suez as the lead operator, through its subsidiary Aguas Argentinas. Suez had also been recently awarded the concession for the city of Johannesburg. Reports had already been prepared by Suez and by the World Bank group that had helped finance the project, but this was the first independent study relating to Buenos Aires' experience with water privatization since their project had begun in 1993. And the South African Municipal Workers' Union, which had convened the press conference, was eager for the people of Joburg to hear the results first hand.

For several years, the Buenos Aires project had been heralded as a major water success story. In 1993, the city's antiquated system of pumps and pipes had had all the makings of a water disaster. Then the Suez-led consortium, after obtaining a 30-year contract, stepped in to modernize and streamline the water delivery system of Latin America's wealthiest city.

Although proponents claim that privatization ensures greater public accountability and transparency, the Buenos Aires project was dictated unilaterally by a presidential decree in 1989. In August 1989, the Argentinian government, under President Carlos Menem, quickly pushed through the National Administrative Reform Law, declaring a state of economic emergency with regard to public services. The law authorized the "partial or total privatization or liquidization of companies, corporations, establishments, or productive properties totally or partially owned by the State . . ." Based on this presidential decree, steps were taken to privatize the Buenos Aires water and sewage network known as Obras Sanitarias de la Nacion (OSN).

According to one study, the consortium did succeed in modernizing OSN's aging infrastructure. Much of the old network was restored and cleaned up and water production increased as a result of basic repairs at one water treatment plant. And by 1999 — six years after the contract was signed — Aguas Argentinas reported that the percentage of the population receiving water service had increased from 70 percent to 82.4 percent, though the Suez-led consortium's five-year target was less, at 81 percent. Despite these positive achievements, however, the Buenos Aires privatization project fell far short of expectations on other fronts.

While privatization, it was hoped, would bring lower water prices for people in Buenos Aires, the net effect has been the opposite. Not long after the Suez-led consortium took over from OSN, a 26.9 percent reduction in water rates was in fact implemented. Yet just after the OSN had been slated for privatization in February 1991, a 25 percent rate increase had been announced, and two months later, prices had gone up another 29 percent. Both increases, it was said, were to compensate for inflation. Further price hikes took place in 1992, and the result was much higher tariffs just before the consortium took over. Taken together, these price increases served to more than offset the initial 26.9 percent price reduction announced at the time of the privatization. What's more, a year after the concession began, the company argued that it needed to increase its prices because the government was making new extra-contractual demands, including a requirement that very poor neighbourhoods receive service immediately. This, the company said, would increase their costs by 15 percent. An additional price hike was granted — 13.5 percent in charges for consumption, disconnection, and reconnection plus a 42 percent increase in an infrastructure surcharge.

The Buenos Aires water privatization plan also allowed Aguas Argentinas to build sewage infrastructure at a slower rate than water infrastructure. By 1999 — one year after the original target date by which the company was to have increased sewage service from 58 percent to 64 percent of the population — the sewage part of the infrastructure program had only reached the 61 percent mark. In the past, the public utility OSN had committed to extending the sewage network at the same rate as the water network. But Aguas Argentina (though it had inherited the disparity between water and sewage services from OSN) maintained that the extension of the water network was more urgent because people in unserviced areas were drinking water polluted by nitrates. It should be noted, however, that removing and treating sewage cost about twice as much as providing water service, while the rates charged for both services were the same. So in the process of providing water services, Aguas Argentina had expanded the more profitable network at a faster rate than the less profitable one. And sewage not collected by Aguas was being disposed of in septic tanks or cesspools or directly into rivers and streams, which led to a risk of health hazards through the spread of water-borne diseases.

Meanwhile, this privatization plan also had its own built-in mechanisms for guaranteeing profits. Indeed, the original contract between the government and Aguas Argentinas had built-in flexibility that did much to protect the company's profit margins. For instance, the agreement allowed Aguas Argentinas to file for a rate increase if its composite cost index (an index based on costs of fuel, labor, and other expenses) rose above 7 percent. And even more flexibility was gained through later contract negotiations. In 1993, performance targets had been set for the first five-year period, but in 1997, partly through contract renegotiation, the five-year period was extended beyond 1998, to the year 2000. A report from the Universidad Argentina de la Empresa stated that by 1995, Aguas

Argentinas was garnering profits of 28.9 percent of revenues, and by 1996 and 1997 these figures were 25.4 percent and 21.4 percent, respectively. These Buenos Aires profit margins were two and a half to three times those of water enterprises in England and Wales — considered by some to be models for privatization — which averaged 9.6 percent in 1998–99 and 9.3 percent in 1999–2000. In short, this water privatization harvested substantial profits.

The root of the problem in Buenos Aires was the fact that a private corporation, whose main goal was to increase profit, was delivering a service that should have been provided by the government on a nonprofit basis. No matter how responsibly a transnational carries out its business, such commercial enterprises are simply not designed, first and foremost, to serve the public interest. Nor are they organized as sustainable enterprises to conserve resources. Since maximizing profits often means encouraging increased consumption, private water delivery enterprises will not work to reduce consumption. Meanwhile, governments are increasingly abandoning their responsibility as guardians of "the commons" — the resources essential to the common good that belong to one and all. Instead, water giants like Suez are taking over, even though their profit-making goals often clash with the needs of the community. In short, "blue gold" is rapidly becoming a big business investment, consolidated by global water markets — "an industry for the 21st century." But who are the main corporate players in this brave new world?

BLUE BONANZA

In a special feature on the global water industry in May 2000, *Fortune* magazine declared: "Water promises to be to the 21st century what oil was to the 20th century: the precious commodity that determines the wealth of nations." This prediction is not surprising, since supplying water to people and industries around the world is already considered to be worth US$400 billion in business annually. And given the fact that water privatization is currently in its infancy, the industry is in a remarkable position compared to other, more established sectors of the global economy. According to *Fortune*'s own analysis, the annual revenues of the water industry amount to approximately 40 percent of the oil sector, and it is already one-third larger than the pharmaceutical sector.

However, industry analysts note that the short-term projections for the water industry are much higher. In 1998, the World Bank predicted that the global trade in water would soon be a US$800 billion industry, and by 2001, this projection had been jacked up to one trillion dollars. This kind of phenomenal growth rate is being projected because the industry's current annual revenues are based on the fact that only 5 percent of the world's population are now receiving their water supplies from corporations. The potential for market growth, in other words, is substantial. At this rate, water could become a multi-*trillion* dollar industry in the future. What if city after city privatizes its water services? *Fortune* conservatively estimates that the industry would then grow at a rate of 10 percent a year, and at the same time, the economic value of water has been escalating. In some parts of the world, reports *Global Water Intelligence* in its monthly analysis of the global water market, water now commands the same price as a barrel of oil. In other places, like the Rocky Mountain Front Range in Colorado, high demand has caused the price of water to triple in one year. Between June 1999 and June 2000, the price of water there shot up from US$4,000 per thousand cubic meters to over US$14,000 per thousand cubic meters.

Given the rising tide of demand for fresh water services in cities around the world, investment strategists are beginning to target the global water industry as "the best sector for the next century." "If you're looking for a safe harbor in stocks," advises *Fortune*'s investment analysts, "a place that promises steady, consistent returns well into the next century, try the ultimate un-Internet play: water." After all, the economic performance of the major corporate players in the industry to date shows that cash flows are stable because they are locked into long-term contracts. As Suez CEO Gérard Mestrallet puts it: "Where else can you find a business that's totally international, where the prices and volumes, unlike steel, rarely go down?"

As the huge U.S. market begins to open up, water corporations are beginning to establish a toehold on Wall Street. In 2000, there were more than US$15 billion worth of acquisitions in the U.S. water industry alone. The *Wall Street Journal* reports that European-based giants Suez and Vivendi are expected to list their stocks on Wall Street markets by the end of 2001. "Water stock fundamentals," says Schwab Capital Markets' analyst Debra Coy, "look good, and over time, the stocks should continue showing gains as they have in recent years, if not outperforming the market . . ." With investors fleeing from declining computer and high-tech stocks, double-digit growth in water shares looks more promising to some. But not all water companies have been enjoying an easy ride on Wall Street. Enron's former water subsidiary, Azurix, is a case in point. In making its debut on Wall Street, says *Global Water Intelligence*, initial expectations about Azurix were high, and the company's stock traded at US$24 a share in June 1999, but "promising a lot and delivering a little," Azurix shares took a nose dive to the point that they were valued at only US$8 the following year. While the North American division of Azurix, now owned by the American Water Works Company, still has a chance to recover its lost stock value on Wall Street, its recent journey indicates that U.S. investors are still getting used to the idea of putting their capital into water corporations.

Online trading, however, could eventually change the way that water companies raise capital. Already, several dot-com companies have been set up to buy and sell water on the Internet. WaterBank.com's website, for example, is designed to provide a "virtual market for water." Other sites like iAqua.com and WaterRightsMarket.com serve as electronic bulletin boards where water buyers and sellers can advertise their products and services. And as *Global Water Intelligence* reports, the Azurix project Water2Water.com has advanced Internet trading by establishing an electronic trading floor that allows buyers and sellers to conduct their transactions directly. A mock trading floor has been set up for the Texas Lower Grande water market as a pilot project.

THE LORDS OF WATER

Today's global water industry is dominated by ten corporate players, which fall into three categories, or tiers. The first tier is composed of the two largest water titans in the world, Vivendi Universal and Suez (formerly Suez-Lyonnaise des Eaux), both based in France. Unlike most countries, which have traditionally entrusted the delivery of water services to their governments, France began to privatize water delivery as early as the middle of the 19th century, under Emperor Napoleon III. As profit-making enterprises, both Suez and Vivendi pioneered the building of the water industry, learning the trade and expanding their operations through their home-based markets. Together, they have monopoly control over 70 percent of the existing world water market. Suez operates in 130 countries and Vivendi in well over 90. While Vivendi is the larger of the two water giants, posting bigger annual sales than its rival mainly because of its diverse operations and large customer base in France, Suez serves far more people (approximately 110 million) around the world. Of the 30 water contracts awarded by big cities since the mid-1990s, 20 went to Suez.

The second tier consists of four corporations or consortiums with water service operations that are (or have been) best positioned to challenge the market monopoly of the two titans: Bouygues-SAUR, RWE-Thames Water, Bechtel-United Utilities, and Enron-Azurix. Based in France, the first contender, Bouygues, currently operates in 80 countries through its water subsidiary, SAUR. The second contender, the German electrical giant RWE, has purchased Thames Water and has thus moved into a position where it could begin to challenge both Suez and Vivendi. Similarly, the partnership between Bechtel, the U.S.-based engineering conglomerate, and United Utilities of the U.K., which provides water services to over 28 million people, could expand both companies' operations. And until its recent divestment of its U.S. water holdings in Azurix, the American energy corporation Enron looked as if it would pose an effective challenge.

The third tier is made up of a group of smaller water companies that have developed considerable capacity and expertise but are not in a position to become the leading corporate players in the global water industry on their own. This tier includes three British companies and one U.S.-based enterprise. The British group consists of Severn Trent, Anglian Water, and the Kelda Group, which was previously known as Yorkshire Water. All three took root after Britain's water system was privatized under Margaret Thatcher in the 1980s. Along with Thames Water and United Utilities, they have cornered the market in the U.K. The fourth company in this tier is the American Water Works Company in the U.S., which recently enlarged its operations by purchasing Azurix.

The corporations that make up the first and second tiers have several other major industrial components, ranging from electricity and gas to construction and entertainment. Only the third-tier companies are focused almost exclusively on water services. Yet they all see themselves as multi-utility providers, their range of expertise as water corporations generally covering four types of services: (1) water and wastewater services; (2) water treatment facilities; (3) water-related construction and engineering; and (4) innovative technologies such as desalination of sea water. In order to build their capacity on these fronts, the corporate players in the water industry have used a series of strategies, including the acquisition of subsidiaries with expertise in these areas, formal partnerships with other companies, and joint ventures with other corporations on specific projects.

At the same time, each of these providers is working internationally to develop a market presence. Suez and Vivendi's transnational operations have already been described; Bouygues services over 25 million people in more than 30 countries through its water subsidiary SAUR; and Enron has held water assets in Mexico, Brazil, and the United Kingdom, as well as the United States and Canada. By purchasing Thames Water, RWE has expanded its operations into the U.K. and Australia, as well as several countries in Asia, the Middle East, Latin America, and parts of eastern Europe. And even though national regulations in the U.K. have curtailed the international operations of the British water companies to some degree, Anglian Water provides water services to over 7.2 million people on five continents while the Kelda Group continues to operate in China, Germany, Canada, and the Netherlands.

These geographical expansions take different forms. In some cases, they involve either public-private partnerships or private joint ventures with other institutions in the region where the water services are to be delivered. In 1999, for example, Vivendi and RWE formed a consortium to take over half of Berlin's water system, the largest privatization ever in Germany's water sector. Another strategy deployed by the big water players is to buy shares in a company already operating in the region, acquire a controlling interest in it, and gradually turn the company into a wholly owned subsidiary. This is what Suez did when, in 1999, it bought the remaining 70 percent ownership of United Water Resources in the U.S.

after initially acquiring 30 percent control in 1994. And some water majors buy up small companies outright to develop new technologies such as R&D programs in water purification and water filtration.

Using some or all of these methods, the global water industry has recently been going through a period of unprecedented growth and expansion. Several common motivating factors are at work in this phenomenon. The constant demand of shareholders for increasing profits and dividends is the prime inducement to grow. It has prompted the water majors not only to expand their marketing operations internationally but also to gain new value-added business opportunities by acquiring more companies. Secondly, in financing the operations of water corporations in the nonindustrialized countries of the South, the World Bank's financing criteria call for the formation of joint ventures or partnerships with other companies where there is a need to strengthen a corporation's multi-utility capacities. Finally, big water corporations are also motivated to grow because of their wide-ranging, international links with governments, political parties, the banking industry, and international financial institutions like the World Bank and the International Monetary Fund.

SUEZ CONQUESTS

Defending the worldwide expansion of Suez into new water markets, CEO Gérard Mestrallet recalls the "philosophy of conquest" that initially formed the foundation stones of Suez as a corporation. Suez was the company that undertook the 19th-century megaproject of building the original Suez Canal, and company founder Ferdinand de Lesseps embraced and advocated this approach as the company's mission. Almost one and a half centuries later, Mestrallet urged that de Lessep's philosophy be resurrected as the corporation's mission in the new global economy. "If we succeed," said Mestrallet, "we shall be in harmony with our world history and follow our culture."

In March 2001, the transnational corporation known as Suez-Lyonnaise des Eaux officially changed its name to Suez. The name change was designed to reflect the corporation's new image as a global multi-utility service provider, and its new structure is based on four core businesses:

water, energy, communications, and waste management services. Most of Suez's EUR$34.6 billion in annual Eurodollar revenues comes from its energy, water, and waste management divisions. Energy — mainly gas and electricity — accounts for 57.4 percent of Suez's revenues and is concentrated in France and Belgium. Water services constitute 26.4 percent of total revenues, with close to three-quarters coming from international markets. Waste management makes up most of the remaining revenues at 14.5 percent, with communications accounting for 1.7 percent.

Water services, however, are clearly Suez's largest growth sector, scoring a 44 percent increase in revenues between 2000 and the previous year. For this reason, Suez consolidated all its water service operations in March 2001 under a new brand name, ONDEO. The new water conglomerate has three divisions: ONDEO Services, specializing in water supply and sanitation; ONDEO Nalco, specializing in water treatment and process chemicals for U.S. industries; and ONDEO Degrémont, specializing in water treatment and turnkey engineering. In making the announcement, Suez boasted that ONDEO would be the "world's strongest and most comprehensive water solutions group."

At the same time, Suez declared that the new ONDEO water company would be "a key step [in] an aggressive growth strategy to increase revenues 60 percent between 1999 and 2004." In addition to their operations in the rest of the European Union, Suez has been securing major water concessions and contracts in Latin America, Asia, and North America — the U.S., for example, being targeted as one of Suez's major growth markets for water services. And in July 2000, Suez had already begun to establish a presence in the United States by purchasing United Water. Operating in 17 U.S. states, United Water will be a key factor in ONDEO's new water marketing plans.

In spite of all these ventures, experience has shown that privatizing public water utilities is of questionable value. For instance, in La Paz, Bolivia, despite receiving a US$40 million loan in 2000 from international financial institutions, Suez was party to a contract that, according to a 1999 World Bank study, did not provide adequate financial incentives for the company to extend water services to some areas. This suggested that service to the poor should be determined by the "ability to pay," rather than as a matter of public policy. Meanwhile, the Public Services International Research Unit (PSIRU) based at the University of Greenwich in the U.K. alleged that Aguas de Limeira, a Suez subsidiary operating in São Paulo, Brazil, had invested only Real 18 million (US$7.2 million) of the Real 36 million (US$14.4 million) that was part of the concession agreement. (According to the company, however, any alleged underinvestment was due to a lack of increase in rates.)

In Europe, the municipal council of Budapest, Hungary, rejected the business plan for water services presented by a consortium of Suez and RWE in July 1999, because they observed that it would result in large losses for the city while reaping huge premiums for management at both companies in the consortium. Plagued by constant wrangling since the contract was first signed, one senior Budapest city official reflected: "It is now clear that this kind of privatization was a mistake."

In the United Kingdom, the government's drinking water inspectorate announced in July 1999 that the Suez subsidiary Northumbrian Water had the second-worst operational performance in England and Wales. The main reason was poor water quality: high levels of iron and manganese were found in the water Northumbrian was delivering. And in Potsdam, Germany, city officials terminated a contract with Suez when the company discovered that water consumption levels were lower than predicted and demanded that much higher water rates be charged. While Suez's actions may have been perfectly legal, a public water system would not need to have sharply increased water rates due to reduced consumption levels, because it would not have been driven by the imperative to maximize profits.

According to the Public Services International Research Unit, Suez's takeover of water contracts and concessions has also sometimes resulted in major worker layoffs. In Manila, PSIRU reported heavy job cuts and selective rehiring — a measure that the corporation would likely defend as a move toward greater efficiency. In Buenos Aires, the water service workforce was cut nearly in half, from 7,600 to 4,000 jobs, when Suez took over, largely through voluntary retirement schemes jointly funded by the government and the Suez-led consortium Aguas Argentina. While Aguas Argentina contends that it has since created thousands of new jobs, studies show that those have been short-term contract jobs (three to six months) with little or no benefits. In Jakarta, Indonesia, water workers went on strike in April 1999, demanding equal pay for all water workers as well as an end to privatized water concessions in the city. In the wake of this, the governor of Jakarta, Mr. Sutiyoso, fired the president of the Water Authority PDAM Jaya. Although transnationals would view these and other wage-saving measures as appropriate means of retaining and increasing profits, such layoffs — as in the case of Jakarta — can result in social disruptions and a lack of continuity in service. This need for uninterrupted service is yet another reason why public utilities like sewage and water delivery operations should remain just that — public.

Like many transnational corporations, Suez is a political machine in its own right. CEO Gérard Mestrallet has held positions with the French government ministries of transport, economy, and finance and has acted as advisor on industrial affairs to the minister of finance. Suez Director Jérôme Monod was chief of staff to ex-Prime Minister Jacques Chirac and currently sits on the board of directors for RWE. Suez's own board of directors includes CEOs and former corporate executives from three major banking institutions in France, a former CEO of Nestlé, and a current director of Shell, as well as Paul Desmarais, Jr., CEO of Power Corporation Canada. Through its U.S. subsidiaries, Suez has also made modest Political Action Committee donations for congressional campaigns and US$141,150 in soft money (unrequested) donations during the 1999-2000 election cycle. Suez is also a key player in the European Forum on Services, the big business lobby machine that pushes for new rules favoring the privatization of public services, including water, at the World Trade Organization.

VIVENDI'S EMPIRE

Few people in North America paid much attention to Vivendi until it merged with Seagrams and Canal+ in December 2000, to form the largest multi-utility service provider of its kind in the world. At this point, the new corporate conglomerate took on a new name, Vivendi Universal, reflecting its status as a fully integrated water, media, energy, telecommunications, and transportation enterprise.

In France, says sociologist Jean-Pierre Joseph, Vivendi Universal has been like an octopus, spreading its tentacles everywhere. In a paper dated January 2001, he wrote: "Picture a teenager from Saint-Etienne or Marseilles, who after drinking a glass of water from the tap, phones a friend . . . then settles down to do his homework, using Nathan or Bordas handbooks and looking up a word in a Larousse dictionary. He . . . then turns off his Bob Marley, Zebda, or Nirvana CD, and for a break, goes to see *Schindler's List* or *Gladiator* at Cinema Pathé. Or he might play computer games like Diablo or Warcraft. At the same time, his father . . . is listening to a concert performed by the Three Tenors, Duke Ellington, [or] . . . U-2, then turns on 'Canal Plus' and . . . connects to AOL (France) to look for work on line . . . [then] he takes out the garbage, to be collected by Onyx. Meanwhile, his wife, who is a doctor, reads some articles in [two medical journals], *Vidal* and *Quotidien* . . . Then she calls a colleague on her portable SFR before helping her young daughter, who is . . . [reading] a book purchased at France Loisirs. This family, in all of their activities, never left Vivendi Universal."

Today, the Vivendi Universal empire is composed of two major divisions, Vivendi Environment and Vivendi Communications. Vivendi Environment, ranked number one worldwide in environmental services, has four subdivisions — water, energy, waste management, and transportation services. Vivendi Communications, ranked number two worldwide in communication and audiovisual services, consists of six subdivisions — television and film, publishing, telecom equipment, and Internet services. In 2000, Vivendi Universal's combined revenues amounted to US$44.9 billion, with Vivendi Environment accounting for close to 60 percent of total revenues. Characteristic of its global reach, 58 percent of Vivendi's revenues are now generated outside of France — 18 percent of those in the U.S. alone. The conglomerate's next largest revenue generators are its water companies — notably, Générale des Eaux (Vivendi's main international water company) and U.S. Filter (the largest water services company based in the United States).

In building its empire, Vivendi Universal is now pinning its hopes on its communications division, connecting phones, television, and computers to high-speed Internet services. "We are starting as the No. 2 global world media company," said Vivendi's CEO, Jean-Marie Messier, following a meeting with investment analysts in New York. "We have one step in front of us to be the No. 1 global company. We want it, we dream it, and we'll do it together." But despite the corporation's steadily rising revenues and profits, investment analysts warn that Vivendi's high debt-to-capital ratio poses a major obstacle to fulfilling these dreams. Vivendi Environment, especially its water consortium, has become the "cash cow" that supports Vivendi Universal. In January 2000, Vivendi unloaded its entire debt onto its environment division and its lucrative water companies, so that its communications division could go debt free. In other words, Vivendi Water holds the key to the empire's future.

Vivendi Universal's marketing strategy is therefore based on privatizing water services and securing water concessions all over the world. And since 1999 alone, Vivendi has successfully acquired an impressive array of long-term water contracts with cities in Asia (Tianjin, China; Inchon, South Korea; Calcutta, India), the Middle East (Tangiers and Tetouan, Morocco; Beirut, Lebanon), eastern Europe (Szeged, Hungary; Prague, Czech Republic), Europe (Berlin, Germany, with RWE), Africa (Nairobi, Kenya; the entire country of Niger; and Chad); and Latin America (Monteria, Colombia). After purchasing U.S. Filter in May 1999, Vivendi also secured a series of concessions in the U.S. and Canada (Onondaga County, New York; Wilsonville, Oregon; Goderich, Ontario; Floyd River, Kentucky; Plymouth, Massachusetts). As the largest water and wastewater company in the U.S. with a market 14 times larger than its nearest competitor, U.S. Filter promises to be a key player in Vivendi's future water expansion plans.

But as in the case of Suez, Vivendi has encountered some difficulties delivering its water services. In 1999, for example, Vivendi's management of Puerto Rico's water authority, PRASA, through its subsidiary Compania de Aguas, was strongly criticized by a Puerto Rican government report in August 1999 for failing to adequately maintain and repair the state's aqueducts and sewers. According to Interpress news agency, "The Puerto Rico Office of the Comptroller [Contralor] issued an extremely critical report on the PRASA-Compania de Aguas contract. The document lists numerous faults, including deficiencies in the maintenance, repair, administration and operation of aqueducts and sewers, and required financial reports that were either late or not submitted at all." The Interpress account of the comptroller's report went on to say, "Citizens asking for help get no answers, and some customers say that they do not receive water, but always receive their bills on time, charging them for water they never get. A local weekly newspaper published reports of PRASA work crews who did not know where to look for the aqueducts and valves that they were supposed to work on." What's more, the 1999 comptroller's report showed that under private administration, PRASA's operational deficit has kept increasing and has now reached US$241 million. As a result, the Government Development Bank (Banco Gubernamental de Fomento) has had to step in several times to provide emergency funding.

In May 2001, the Puerto Rico Office of the Comptroller issued another report about PRASA's performance, identifying 3,181 deficiencies in the administration, operation, and maintenance of the water infrastructure. Among these, the Comptroller reports that PRASA's operating losses had increased from US$241 million in August 1999 to US$695 million in May 2001, and that the agency had not collected US$165 million in bills. The report also noted that the U.S. Environmental Protection Agency had fined PRASA a total of US$6.2 million since it had been privatized through Vivendi's Compania de Aguas (i.e., during the period between 1995 and 2000). According to Comptroller Manuel Diaz-Saldaña, the privatization "has been a bad business deal for the people of Puerto Rico." "We cannot keep administrating the Authority (i.e., PRASA) the way it has been done until now," he said.

Elsewhere, Vivendi took the Government of Argentina before the International Centre for Settlement of Investment Disputes (ICSID), a division of the World Bank, claiming that it had violated a Bilateral Investment Treaty by not preventing the City of Tucumán from taking action against the company over its water contract. Tucumán officials had charged Vivendi for poor performance in the running of its water system, citing multiple cases of brown water. The court dismissed the claim, saying there was no evidence that "the Argentine Republic failed to respond to the situation in Tucumán and the requests of Vivendi in accordance with the obligations of the Argentine government under the BIT " (the Bilateral Investment Treaty between France and Argentina).

Meanwhile, Vivendi's joint venture with Sereuca Space to manage the water billing and revenue system for Nairobi, Kenya, has become the subject of a major public controversy. Peter Munaita, writing in the *East African*, reported in August 2000 that Sereuca Space, in a joint venture with Vivendi's subsidiary Générale des Eaux and Israel's Tandiran Information Systems, "will not invest a single cent in new water reservoirs or distribution systems during the ten years the contract will be in force. Instead, the company will spend an undisclosed amount on installing a new billing system at City Hall and, for that, reap 14.9 per cent of the [estimated] Ksh12.7 billion [$169 million] collected over the period." Says Munaita, "Nairobi deputy mayor, Mr. Joe Aketch, has opposed the deal, saying it will lead to a loss of 3,500 jobs in exchange for 45 staff, four of them expatriate, who Sereuca proposes to employ."

In response to widespread public criticism of the proposed project, Vivendi said later it would invest another $150 million in expansion, repair, and maintenance to minimize water loss. In August 2001, however, the Kenyan government announced that it was suspending the water billing project until the World Bank had completed a privatization option study. According to World Bank officials, the proposed water billing contract was too expensive and was not tendered for through commercial bidding procedures. Says World Bank officer Peter Warutere, "The study will achieve cheaper alternatives." According to reports by Munaita, Vivendi maintains the contract's suspension will "jeopardise prospects of Nairobi getting a reliable water service before 2008, with shortages becoming rampant in two year's [sic] time," because "World Bank projects take between four and seven years before implementation starts."

In Berlin, the German Green Party launched a court challenge claiming that Vivendi's water prices and guaranteed dividends (assured 15% profit regardless of productivity) were unconstitutional. The Constitutional Tribunal court agreed. Vivendi responded that they would renegotiate the contract to put themselves in accordance with the court's judgment. And in the United Kingdom, a joint venture between Vivendi, Suez, and Bouygues was publicly condemned when it laid off 3,200 workers after the British government ordered the consortium to reduce its water prices.

In attempts to guarantee an ever-expanding market for its services, including water, Vivendi Universal has been busy carving out a political role for itself, promoting a new set of global rules for cross-border trade in services. Vivendi is one of very few transnational corporations in the world that sits on both of the two most powerful big business lobby groups — the U.S. Coalition of Service Industries and the European Forum on Services — currently involved in the negotiation of the General Agreement on Trade in Services (GATS) at the World Trade Organization. (See Chapter 7.) Vivendi's CEO, Jean-Marie Messier, has also been playing a leading role in building a new consensus among governments and corporations concerning a set of global rules for the promotion of electronic commerce on the Internet. And the political links have been strengthened by the fact that Vivendi's board of directors includes several high-profile business leaders with significant political connections like Dick Brown of Electronic Data Systems (EDS) (which boasts former U.S. Secretary of Commerce William Daly on its board). During the 1999–2000 election cycle, various U.S. components of the Vivendi empire — including Universal Studios, U.S. Filter, and Philadelphia Suburban — contributed a total of US$186,000 to Political Action Committees and another US$40,110 in soft money donations.

ENRON'S GAMBLE

At the dawn of the 21st century, the potentially lucrative global water market was monopolized by two France-based giants, Suez and Vivendi. Between them, they had captured over 70 percent of the worldwide market, with operations in over 130 countries. The market, however, was still in its very early stages, so the question on the minds of many market analysts was "Who will break their stranglehold?" Most predicted that it would be the corporations in tier two of the global water industry. After all, these were the enterprises with the capital and global market reach needed to mount an effective challenge to Suez and Vivendi. But to do so, any contender from tier two would need to fortify itself with the more specialized skills and experience of the water companies in tier three. Enter Enron, the global energy services giant, accompanied by newly formed water company Azurix.

Enron had been growing at a spectacular rate. Its online energy marketing system had become the largest e-commerce website in the world, and its wholesale energy services division was delivering twice the natural gas and power volumes of its closest competitor. Enron Transportation Services had been formed to carry on with the company's gas pipeline operations while Enron Energy Services became the retail arm focusing on commercial and industrial users of energy. With this growth, Enron's revenues reached record levels. Between 1999 and 2000 alone, Enron's total revenues increased by a whopping 151.3 percent, from US$40.1 billion to US$100.8 billion. During this period, the corporation's electricity sales doubled while its sales in natural gas rose by a third — most of these gains resulting from the California energy crisis which Enron was well positioned to exploit. By 2000, Enron's total revenues were more than those of Suez and Vivendi Universal combined, and the company was not saddled with a huge debt load.

In 1998, Enron, according to one of its own press releases, made its bid to "exploit the worldwide move to the privatization of water." After the company purchased Wessex Water of the U.K., the stage was set to establish Azurix as a subsidiary that could become a major player in water and wastewater services. Rebecca Mark, an emerging star in the Enron ranks, was made president and CEO of Azurix. Once declaring that she "would not rest until all the world's water had been privatized," Mark turned Azurix into a company that could provide a wide range of business services such as managing municipal water delivery, constructing water plants, developing wastewater distribution systems, and disposing wastewater treatment by-products. Building on Wessex Water's experience, Azurix acquired water companies or concessions in Argentina, India, Bolivia, Mexico, and Canada, and it formed a joint venture in Brazil. In 1999, for example, Azurix took over the Canadian-based Philips Utilities, which manages a variety of water and wastewater projects in the United States and Canada. As all these initiatives were being carried out, Enron's own marketing expertise and contacts in the electricity and natural gas industry helped the company carve out a niche in the global water market.

Enron's extraordinary political connections were also a key asset. During the former presidencies of George Bush and Bill Clinton, Enron's political reach into the White House was well known in Washington circles, and with the election of George W. Bush, the political ties appear to have become even stronger. Enron's CEO, Kenneth Lay, is especially well placed. During the 2000 presidential campaign, Lay was part of Bush's Pioneer Group, composed of the four hundred or so people who had personally contributed US$100,000 or more to his election drive. Certainly, Kenneth Lay has been a key player on Bush's Energy Advisory Panel and Vice-President Dick Cheney's newly formed Energy Policy Development Group. Enron has also made substantial cash donations, including US$300,000 for George Bush's inauguration party and a total of US$2,387,848 for candidates during the 2000 election cycle. Beyond this, Enron has played an influential role in the major big business lobby networks such as the U.S. Coalition of Service Industries, the board of the National Foreign Trade Council, and the U.S. Council for International Business.

Yet even with all of Enron's economic and political clout as its parent corporation, Azurix was unable to make the breakthrough it needed to become a major player in the global water market. From the outset, Azurix had difficulty competing effectively in bidding wars with Suez, Vivendi, and other global players for lucrative water concessions. Expected to be the main revenue source for Azurix, Wessex Water's performance turned out to be disappointing. The company's revenues declined after a cap was placed on water rates in April 2000 by the Office of Water Services (Ofwat), the water-industry regulator in England and Wales. At the end of 1999, the price of Azurix shares plunged 40 percent in a single day and never recovered. On several occasions, Enron stepped in to bail Azurix out with loans. But following a court battle with the water company's shareholders, Enron announced on December 21, 2000, that a deal had been reached to buy back Azurix's stock market shares for US$325 million, and this set the stage for Enron to evaluate its future role in the water industry.

In its short experience, Azurix had encountered numerous setbacks, notably with its water concession in Bahia Blanca, a city 420 miles southwest of Buenos Aires in Argentina. During the year 2000, residents launched numerous complaints about poor water quality and low water pressure in the city's system, which was being managed by Azurix. Early that year, authorities had warned residents that their tap water was contaminated with bacteria due to an algae outbreak in the city's reservoir. For months, the water had a bad odor and taste. Said public health chief Ana Maria

Reimers: "I've worked here for 25 years and I'd say this is the worst water crisis I've ever seen here." "The situation is not of Azurix's making," declared Richard Lacey, Azurix's managing director of technical operations, in May 2000. "It's a result of the poor quality of water supplied by the provincial government's reservoir and dam."

When a public water system is privatized, the chain of responsibility becomes exceedingly murky. A water concession, like that of Azurix in Bahia Blanca, working at arm's-length from the government, becomes much more difficult to regulate directly than a public water agency. In January 2001, it was reported that Provincial Governor Carlos Ruckauf was going to ask provincial legislators to consider canceling the 30-year contract with Azurix altogether. But later news reports showed that the Public Works Minister, Julian Dominiquez, preferred to negotiate improvements with the concession rather than seeking cancellation. Then in February 2001, Azurix agreed to spend $30 million on improving its water and sewage services in response to the public complaints. But in July 2001, Azurix wrote to the provincial government of Buenos Aires saying that the concession was not economically feasible, and in September 2001, Azurix's Latin American CEO, John Garrison, met with Governor Ruckauf and Minister Dominiquez to lay the groundwork for the company's withdrawal from the concession. At the same time, it was reported by the daily *El Dia* newspaper that the company would sue the province of Buenos Aires for a sum of up to $400 million.

Meanwhile, in April 2001, Enron had announced that its troubled water company, Azurix, would be broken up and its assets sold. Enron's gamble with Azurix had not paid off. The energy services giant had become "frustrated with the water sector," reported *Global Water Intelligence*, and did not have the "patience" required to build a water company capable of competing for markets in this industry over the long haul. Instead, according to this article, Enron had become too accustomed to "the kind of high velocity money" that the energy sector had been delivering recently. Four months later, American Water Works Company announced that it had bought the assets of Azurix in North America. With this acquisition, American Water Works would strengthen its water market presence in the southeastern and northwestern U.S. and in three Canadian provinces.

Yet Enron's breakup of Azurix merely foreshadowed its own dramatic collapse. Eight months later, the global energy services giant was filing for bankruptcy protection, as creditors, regulators, and politicians moved rapidly to tighten the noose around the debt-ridden Enron. Instead of being the financially healthy, rapidly rising corporate star on the Global Fortune 500, Enron suddenly faced a mountain of debt totaling at least $13 billion in December 2001. As the U.S. Securities and Exchange Commission stepped up its investigation of Enron's accounting practices and potential conflicts of interest, the energy colossus that had spearheaded the drive toward the privatization and deregulation of public services was now becoming known for having filed the biggest bankruptcy claim in history.

NEW CONTENDERS

Enron will not be the last of the challengers to enter the world water market. The prospects of a multitrillion-dollar blue bonanza is too great to be left in the hands of a few corporate conglomerates, and there are signs that new contenders are emerging. Some of them may, in the coming years, be in a position to take on the water empires of Suez and Vivendi. Once again, a series of acquisitions and/or mergers are in the works which, if successful, could break up the global water market monopoly and unleash more forces of privatization. Two that have recently begun to surface involve German-based corporate players.

The first has to do with the multi-utility giant RWE, which acquired Thames Water to establish its global water market base. In Germany, RWE is currently the second-largest electricity company and one of the largest waste management operators. With annual revenues averaging well over US$40 billion, RWE has consistently ranked high in the Global Fortune 500, and recently, RWE has been trying to restructure its operations to become a multi-utility corporation supplying energy, water, waste management, and telecommunication services to urban centers around the world. In the water delivery market, RWE had begun to enter the big leagues through its joint ventures with Suez in Hungary and Vivendi in Berlin. And its September 2000 takeover of Thames Water, then the leading third-tier player, was meant to consolidate RWE's international presence in the water sector. "Thames gives us the scale and the technical expertise to become a worldwide player," said RWE's CEO, Dietmar Kuhnt.

A year after the takeover, RWE was registering a 29 percent increase in total revenues, at EUR$62 billion, and a 35 percent increase in operating profits. RWE's water division, based on the strength of Thames, accounted for 20 percent of the profit increase, and during this first year in the RWE fold, Thames had already begun to extend its market reach. After setting up a water treatment plant at Shanghai in 1995 — the first foreign-based company to set up such an operation in China — Thames secured a contract to jointly manage the city's water supply system with the state-owned Pudong Tap Water Co. in 2001. In Thailand, Thames won a US$240 million contract to supply water services in two provinces, the largest Asian water concession signed to date in 2001. And shortly after the RWE takeover, Thames secured a controlling interest in a Chilean water and sanitation company, ESSBIO, which provides services for the 1.5 million people in Chile's second-largest city, Concepcíon.

In spite of these corporate gains, Thames Water's performance record has drawn negative public comment in the United Kingdom. On July 27, 2001, U.K. Environment Minister Michael Meaker said, "I am extremely worried by Thames Water's performance. Its inability to cut leakage, or even to account for where all the water in its pipes goes, is totally unacceptable. I fully support the firm action that [the U.K. water-industry regulator] Ofwat has proposed, and Thames has accepted, to rectify this situation." Between April 1999 and April 2000, Thames lost enough water to fill three hundred Olympic-size swimming pools a day, according to Ofwat. In August 2001, Thames pleaded guilty in court and was fined GBP26,600 for allowing raw sewage to pollute a stream located within yards of houses in a British community.

In September 2001, RWE took another step in fortifying its position as a new contender. RWE bought the American Water Works Company and its water service operations in the U.S. This included the Azurix holdings in the U.S. recently sold by Enron.

The German energy giant E.ON is another contender in the battle to build an effective counterweight to the corporations that currently have a stranglehold on the world water market. To diversify its operations as a multi-utility corporation, E.ON went on a shopping spree in 2000, looking for a water company to buy so that it could become a player in the growing business of private water distribution. After initial talks with Suez and Enron failed to produce a deal, E.ON made a bid to buy SAUR, the water company wholly owned by the French construction and telecommunications enterprise Bouygues. After Vivendi and Suez, SAUR is recognized as the next-largest water services company in the world, albeit a distant third. But with the backing of a conglomerate like E.ON, which had extensive capital to invest, SAUR could get the boost it needed to make significant gains. To date, however, the E.ON bid for SAUR has not been accepted by Bouygues.

By 1999, SAUR had already established operations in as many as 80 countries around the world and especially in Latin America. In September 2000, SAUR joined with the Spanish water company Aguas de Valencia and formed a new consortium to open new markets in Latin America for water privatization. A year later, reports indicated that SAUR was negotiating with Enron to pick up the remaining Buenos Aires assets of Azurix. At the same time, SAUR secured a contract to become the main operator of the water supply and treatment facility for the country of Mali in northeast Africa. The Mali contract adds to a string of water and electricity operations that SAUR International has already opened up in Africa, including operations on the Ivory Coast and in Senegal, Guinea, the Central African Republic, Mozambique, and South Africa. And in Poland, SAUR beat out Vivendi for a 25-year contract to manage and modernize the water and drainage system in Ruda Slaska.

Regardless of whether RWE-Thames or E.ON or any other alliance of corporate players from tiers two and three are able to mount an effective challenge to Suez and Vivendi, the competing forces will undoubtedly unleash increasing waves of privatization. Unless these corporate armies of privatization are checkmated, water as the essential ingredient of life itself could easily become almost completely commodified and commercialized by the end of the first decade of the 21st century.

PRIVATIZED FIASCO

Continued widespread privatization of water will also be a recipe for an inequitable and nonsustainable future. The model of privatization presents a disturbing picture as the commons and society are being assaulted from new angles. In particular, we need to take a closer look at the track record of the global water corporations and their implications for labor, quality of life, and the environment.

In making his pitch on privatization at an industry conference in 1997, Jeffrey Skilling, President of Enron, provided the following advice: "You must cut costs ruthlessly by 50 to 60 percent. Depopulate. Get rid of people. They gum up the works." Notwithstanding Skilling's blunt rhetoric, his words reflect the philosophy that motivates many transnationals as they seek substantial profits. Since the name of the game is to maximize those profits, cutting costs means laying off workers while raising water rates to generate more revenues.

Suez's takeover of water concessions in Manila and Buenos Aires, as well as Enron's labor difficulties in Argentina and elsewhere, illustrate these methods of privatization. In Manila, Suez and United Utilities promptly cut jobs in water services when they took over the concession. In Buenos Aires, the water workforce was slashed from 7,600 to 4,000 after Suez took over. Enron has also actively opposed union positions in the U.K., Argentina, Guatemala, and India. Unless governments negotiate strict and enforceable conditions regarding job protection as in the Berlin water concession, then it is predictable that jobs will be lost and workers' rights placed in jeopardy. Indeed, many privatization plans often lead to large-scale layoffs as corporations focus on their main goal — increasing financial returns to their investors.

Perhaps more disturbing is the health and safety record of some of the major corporate players in the world's water markets. In 1985, for example, Bechtel was fined by the U.S. Nuclear Regulatory Commission's Office of Investigations for deliberately circumventing safety procedures during cleanup operations at the Three Mile Island nuclear reactor in Pennsylvania. And in 1995, Enron was fined by the U.S. Occupational Safety and Health Administration for numerous safety violations pertaining to the 1994 explosion that occurred at its methanol plant in Pasadena, Texas.

These are examples of past violations of health and safety standards in other sectors of industrial activity, but they are still cause for concern, since these same corporations are involved in, or plan to be connected with, water delivery operations.

Similarly, the environmental track record of many corporate players demonstrates that privatized water management is itself nonsustainable. In the United Kingdom, the U.K.'s Environment Agency has cited many major water companies as being among the worst environmental offenders in the country. Between 1989 and 1997, Anglian, Northumbrian, Severn Trent, Wessex, and Yorkshire Water were successfully prosecuted 128 times for violations ranging from water leakages to illegal sewage disposals. And between 1990 and 1997, according to the U.S. Environmental Protection Agency, Bechtel is reported to have been responsible for 730 spills of hazardous materials while Enron was listed as responsible for 76 spills, some of which were very large.

The model of privatization itself creates enormous disparities in power between corporations and the local governments that usually deal with them. For the most part, water concessions involve a transfer of concentrated power into the hands of private corporations. As a result, governmental power is greatly reduced, making it difficult, if not impossible, for them to establish minimum access and quality requirements. Nor can governments always be effective in penalizing corporations for failing to meet water quality standards while continuing to raise water rates. In one instance, Britain's water regulator, Ofwat, responded to public complaints about rate increases and poor water quality by requiring water companies with U.K. operations to reduce their water rates and improve water infrastructure. But Suez, like other global water corporations, had a considerable amount of maneuvering room to offset this measure. It announced that it would slow down its environmental investment programs and it would not comply with a schedule set down by the E.U. to adopt certain environmental standards.

Instead of facilitating greater efficiencies and ensuring equitable distribution, the model of privatization is designed primarily to enhance corporate profits. Another U.K. example illustrates this point: the Suez subsidiary Northumbrian Water. Between 1989 and 1995, Northumbrian's water rates increased by 110 percent; the CEO's salary increased by 150 percent; and the company's profits increased by 800 percent. The built-in fallacy of this model is that it is ultimately nonsustainable. It demands increasing consumption while contributing little to the conservation of resources.

Meanwhile, the landscape of the global water industry has been marred by a number of legal problems. One of the more celebrated cases involves Suez and the city of Grenoble in France. After an investigation into allegations of corruption, a team of magistrates concluded that Grenoble's water service had been privatized in 1989 in exchange for donations totaling 19 million francs, made by Suez-Lyonnaise des Eaux to the election campaign of the city's mayor, Alain Caigon. In 1996, both Caigon (who by then was minister of communications in the central government of France) and Jean-Jacques Prompsey (who by then was chief executive of Suez's international waste management division) were convicted of accepting/paying bribes and sentenced to time in prison. Subsequently, the courts also ruled that the citizens of Grenoble had been damaged by the corrupt deal and gave them the right to claim compensation.

Another case of bribery had to do with Vivendi and the city of Angoulême in France. In 1997, Jean-Michel Boucheron, former mayor of Angoulême (and later a junior Cabinet minister in the central government) was convicted and sentenced to two years in jail (plus two suspended years) for taking bribes from companies that were bidding for public service concessions in Angoulême. In yet another case, reports David Hall of the Public Services International Research Unit (PSIRU), executives of Générale des Eaux were convicted of bribing the mayor of St-Denis on the Ile de la Réunion in France in order to obtain a water contract. And in 1998, the Oregon Court of Appeal ruled that Enron's subsidiary Portland General Electric had been overbilling its customers US$21 million a year.

Finally, the construction units of Bouygues, Suez, and Vivendi have all been the subject of a major judicial investigation in France over allegations that they participated in a corrupt cartel between 1989 and 1996. According to PSIRU reports, the three corporations are alleged to have shared contracts worth approximately US$500 million for the building of schools in the Ile de France region surrounding Paris, to the exclusion of other bidders. In addition, it has been alleged that a 2 percent levy was to be charged on all contracts for use in support of political parties in the region. This arrangement (assuming it is proven) was described in *Le Monde* as "an agreed system for the misappropriation of public funds."

~~

Though supporters of privatization have claimed, incredibly, that for-profit corporations are likely to be more publicly accountable and transparent than elected governments, the exact opposite tends to be the case. Transnational water delivery enterprises around the world are making substantial profits while water rates have risen in many privatized jurisdictions — too frequently out of the reach of the poor. This has come about because the main goal of a private enterprise is not to serve the public or to make sure water is distributed equally to all users whether at profit or not. Its main goal is to serve its shareholders — to increase profit for a select few. And the concentration of power represented by the huge profits to be gained from the worldwide water business has resulted in some company officials misusing that power.

In effect, "corruption is a systemic feature of privatization processes," concludes PSIRU, "in water as in other areas." Furthermore, this view is confirmed by no less than the World Bank itself in a report called *The Political Economy of Corruption: Causes and Consequences*, which states: ". . . the privatisation process itself can create corrupt incentives. A firm may pay to be included in the list of qualified bidders or to restrict their number. It may pay to obtain a low assessment of the public property to be leased or sold off, or to be favoured in the selection process." The World Bank report goes on to say: ". . . firms that make payoffs may expect not only to win the contract or the privatisation auction, but also to obtain sufficient subsidies, monopoly benefits, and regulatory laxness in the future."

One of the operating assumptions behind the privatization agenda is that public service providers are inefficient. While this may be true in some cases, it is by no means universal. Take, for example, the case of Chile. Since 1998, most of the public water services in Chile have been partially privatized, mainly through the sale of shares to private water giants. Yet prior to this privatization, the public water utilities were recognized as being highly efficient operations. In a 1996 comparative study of six developing countries, the World Bank underscored Chile's public water companies, especially EMOS, as model examples of efficiency. As the Public Services International Research Unit points out, financially efficient public water undertakings provide profitable opportunities for the private sector. "Where one of the motives is raising money to finance the public authority's budget," says David Hall of PSIRU, "then there is in fact a perverse motive to privatize the most efficient water undertakings, because they will obtain a higher price."

What's more, the world's largest public water utility, SABESP, in the state of São Paulo, Brazil, has undergone extensive restructuring since 1995 to make its operations more modern and efficient. Serving the majority of the state's 22 million inhabitants, SABESP has been reorganized in such a way as to expand its revenue generation on the one hand and to cut excessive costs and inefficiencies on the other. In 1995 alone, reports PSIRU, "the population in the service area supplied with treated water increased from 84% to 91%, the population receiving sewage services increased from 64% to 73%, and non-functioning accounts plunged to 8%." Overall, the operating costs of the public water utility were reduced by 45% and SABESP is now in a position to finance its investment programs through loans and its own funds (although the Brazilian currency devaluation of 1999 did have a negative impact on the company's finance capacities). At the same time, SABESP has expanded its environmental responsibilities, including participation in the cleanup of the Tiett River, considered to be the largest environmental project of its kind in Latin America.

In spite of the efficiencies of public utilities such as this, however, the privatization juggernaut is rolling on, and the global water lords are not only taking over local water distribution systems. They are also lining up to conquer the domain of bulk water export.

6

EMERGENT WATER CARTEL

How corporations and governments
are poised to mobilize global trade in water

In February 1999, Terence Corcoran, editorial writer for the *National Post* newspaper in Canada, caused a buzz in international business and government circles when he predicted that there would be an OPEC of Water by the year 2010 and that Canada would be at the center of it. In ten years, he wrote, "Canada will be exporting large quantities of freshwater to the U.S., and more by tanker to parched nations all over the globe." Corcoran predicted that just as OPEC had become the global cartel of the major oil-rich nations of the Middle East, the water-rich countries, including Canada, would band together over the next decade in an "attempt to cartelize the world supply of water and drive the price up." Moreover, wrote Corcoran, once governments wake up to the fact that there is big money to be extracted from massive water shipments, "Canada will end up scrambling to head the WWET, the World Water Export Treaty, signed in 2006 by 25 countries with vast water reserves."

Although Corcoran's predictions were publicly denounced as being far-fetched, a closer look at the global water industry reveals that corporations have been working with governments since the early 1990s on plans to develop massive schemes for bulk water exports from water-rich regions of the world. And in 1996, the World Bank's leading expert for North Africa and the Middle East declared that one way or another, "water will be moved around the world as oil is now." "Within the next five years," he said at the time, "we'll see a rising recognition that water is an inter-national commodity." And back in 1991, the *Globe and Mail* newspaper's *Report on Business* magazine had predicted that "pollution, population growth and environmental crusading are expected to put enormous pressure on the world's supply of freshwater over the next ten years." "Some of Canada's largest engineering companies," the commentary continued, "are gearing up for the day when water is moved around the world like oil or wheat or wood . . . What will be important is who has the right to sell it to the highest bidder."

By the dawn of the 21st century, a whole new set of corporate players was on the scene promoting plans for worldwide water exports. As one of the new contenders, Global Water Corporation (recently renamed Global H2O), put it: "Water has moved from being an endless commodity that may be taken for granted to a rationed necessity that may be taken by force." For the corporate players, it all boils down to supply and demand factors when it comes to the issue of bulk water transfers within the global market system. On the supply side, there are the water-rich countries or regions of the planet that have an abundance of stored fresh water in the form of lakes, rivers, and glaciers: Alaska, Canada, Norway, Brazil, Russia, Austria, and Malaysia, among others. On the demand side are the water-poor countries or regions of the planet where there is a scarcity of fresh water reserves because of desert conditions, dried-up aquifers, or water contamination. They include the Middle East, China, California, Mexico, Singapore, North Africa, and many other countries and regions on virtually every continent. The name of the game is to secure control over bulk water supplies and deliver them to targeted demand areas based on the "ability to pay," at a price that will not only cover costs but also satisfy the desire for increasing profit margins.

During the 1990s, new technologies were developed for transporting bulk water supplies to market through water bags and bottled water, as well as pipelines, supertankers, and canals. Although it is argued that bulk water exports will be too expensive to be economically viable, the World Bank warns that all the low-cost, easily accessible water reserves are already tapped. New water supplies, however developed, will be two to three times more expensive. Yet, says the World Bank, the demand will be there regardless of the cost. Nor is desalination of sea water likely to be an adequate substitute for bulk water exports. While some countries will make use of desalination, it is a very expensive and fuel-intensive process. Massive desalination projects would be an option only for countries that have abundant energy supplies, and it would seriously add to global warming — a crisis already exacerbated by fresh water diversion.

Yet bulk water exports also pose a serious ecological threat. While more impact studies are necessary, there is sufficient evidence that draining bulk water from lake and river basins disrupts ecosystems, damages natural habitat, reduces biodiversity, and dries up aquifers and underground water systems. The damage is even more extensive when water is transported in bulk form to desert regions that were never meant to support human habitat in large numbers. Take the case of the Arizona desert, where the population has multiplied tenfold in the last 70 years to four million, with more than eight hundred thousand in Tucson alone. In an *Atlantic Monthly* article on "Desert Politics," writer Robert Kaplan captures the dilemma in dramatic terms:

> *Maybe, as some visionary engineers think, the Southwest's salvation will come ultimately from that shivery vastness of wet, green sponge to the north: Canada. In this scenario a network of new dams, reservoirs, and tunnels would supply water from the Yukon and British Columbia to the Mexican border, while a giant canal would bring desalinated Hudson Bay water from Quebec to the American Midwest, and super tankers would carry glacial water from the British Columbian coast to southern California — all to support an enlarged network of post-urban, multi-ethnic pods pulsing with economic activity.*

Kaplan's "visionary engineers" may be further along than we think. In response to the so-called "demand" of water-scarce regions ranging from Arizona, California, and Mexico to China and Singapore, and from the Middle East and northern Africa to Spain and Greece, multibillion-dollar megaprojects for bulk water transfers are being planned and implemented.

PIPELINE CORRIDORS

Pipelines, of course, have long been used for irrigation purposes in agriculture. But now new pipeline technologies are being developed for bulk transfers of water on a cross-continental basis. In Europe, for example, a high-tech pipeline has been constructed to transport spring water from the Austrian Alps to Vienna. Now plans are in the works to extend this pipeline to other countries through the proposed European Water Network. Over the next decade, the plan is to construct a pipeline corridor to transport alpine water from Austria into Spain and Greece. For Austrian environmentalists, however, this water pipeline project has caused unease and they now warn about the damage that bulk exports could have on the sensitive ecosystem that exists in alpine regions.

In Turkey, government and corporations are also turning to pipeline corridors as a means of transporting bulk water supplies to markets. Pipelines, as well as converted oil tankers, will be used to transfer water from the Manavgat River in Turkey to markets in Cyprus, Malta, Libya, Israel, Greece, and Egypt. In the summer of 2000, Israel already began negotiations to buy over 13 billion US gallons (49 billion liters) of water a year from Turkey. Turkey's water company says that it has the pumps and pipes in place to export four to eight times that amount. As a result, Turkish corporations are poised to be major sellers and exporters of bulk water supplies through pipelines to thirsty regions of central Europe.

In the U.K., British and Scottish corporations have been exploring the prospects of large-scale exports of water from Scotland, by pipeline and tanker, in response to the growing crisis of fresh water scarcity in England. According to Strathclyde University professor George Flemming, it would be relatively simple to build on existing infrastructure to create a pipeline corridor from the north of Scotland and Edinburgh to London and other parts of England. But support for water sovereignty is strong in Scotland, so when Scotland's water authority, West of Scotland Water, publicly sounded out a plan to sell surplus water to Spain, Morocco, and the Middle East, it was forced to back off because of public reaction. Yet observers maintain that bulk water exports from Scotland by pipeline are inevitable, given the extent of predicted water shortages in England and Wales.

At the same time, United Water International has secured the water concession for Adelaide in southern Australia and is also developing plans for bulk water exports. (United Water International is jointly owned by Vivendi, RWE-Thames Water, and Brown & Root, an engineering firm.) The 15-year strategy is designed to export water to other countries for use in computer software manufacturing and agribusiness irrigation, using a combination of pipelines and tankers for transportation. Initially, local Australian companies were not even allowed to bid for the contract for bulk water exports because it was assumed that an international consortium would be needed in order to increase the value of exports, now expected to be in the range of $628 million.

One of the more notorious pipeline schemes, as we saw in Chapter 1, is Colonel Moammar Gadhafi's multibillion-dollar project in Libya to mine the underground aquifers of the Kufra Basin in the Sahara Desert for bulk water transfers to other parts of the country. One stage of the project, for example, involves two 4-meter (13-foot) pipelines: one carrying up to 700 million cubic meters (about 24.7 billion cubic feet) a year to coastal farms and the other transporting 175 million cubic meters (about 6.2 billion cubic feet) a year to communities along the northwestern mountain range. The project is being built by the South Korean construction and transportation conglomerate Dung Ah Construction Industrial Company, headed by Choi Won Suk, known as the "Thinking Bulldozer" or "Big Man" for his approach to mammoth projects like this.

Plans are being made in North America as well for constructing a network of pipeline corridors to transport bulk water from rivers, lakes, and glaciers in the Canadian North and Alaska to California and other thirsty regions in the United States. Speculation has it that President George W. Bush's call for the construction of a massive energy pipeline corridor from Alaska and the Canadian North may not be about transporting oil and gas alone, but also about exporting bulk water. And questions are already being raised by farmers in Alberta and Saskatchewan about whether the thousands of miles of new pipeline currently being laid across their farmlands will be part of a water-exporting network.

SUPERTANKERS

The call to export bulk water by supertanker has been heating up, especially in North America. Such shipments are sent in huge supertankers, originally designed to ship oil, and if this enterprise became widespread, some shipping companies would transport both oil and water. As Canadian water specialist Richard Bocking explains, their tankers would empty oil on one leg of the trip and carry water home on the return voyage. The first tanker shipment of water out of the United States, says the assistant general manager of Anchorage Water and Wastewater Utility in Alaska, may have been in a tanker leased by the Japanese trading conglomerate Mitsubishi. In 1995, this Mitsubishi-leased tanker shipping petroleum by-products overseas loaded a couple million gallons of water from Eklutna, Alaska, for transport back to Japan.

Along the Pacific coast, supertankers shipping bulk water would likely operate year round on tight schedules, moving through treacherous seas and leaving serious ecological damage in their wake. "These huge tankers," says Bocking, "would wind their way through tortuous coastal waterways, maneuvering around islands and reefs in an area where no well-developed marine traffic management system exists . . . Pods of killer whales move regularly through these waters. Along with commercial and sports fisheries, spawning for almost the entire commercial oyster industry of coastal B.C. is located here." For Bocking, the danger lies in the fact that the massive fuel tanks of these supertankers "are full of bunker C fuel, the worst possible grade of oil in environmental terms. With currents, winds, rocks, and reefs intersecting with tight ship schedules, the stage is set for tragedy on a grand scale."

Before the British Columbia government banned bulk water exports in 1993, a number of companies had been formed with plans to transport water by supertanker along the Pacific coast, including Western Canada

Water, Snow Cap Water, White Bear Water, and Multinational Resources. One project was to involve a Texas company prepared to pay for a fleet of 12 to 16 of the world's largest supertankers (500,000 deadweight tons) to operate around the clock. Under one contract, the annual volume to be shipped to California was equivalent to the total annual water consumption of the City of Vancouver in Canada. Now that there has been a change of government in British Columbia, there are fears that the export ban may be reversed, thereby opening the floodgates for bulk water shipments by supertanker down the Pacific coast.

Meanwhile, Alaska was the first jurisdiction in the world to permit the commercial export of bulk water. Alaska's potential for water exports is reported to be staggering. In Sitka, Alaska, says the pro-export *Alaska Business Monthly*, a one-million-gallon tanker could be filled every day and this would still represent less than 10 percent of the region's current water usage. In Eklutna, Alaska, it is estimated that the export potential could be as high as 30 million US gallons (about 113 million liters) per day. "Everyone agrees water has 21st century potential as an export from Alaska," says the *Alaska Business Monthly*, "and communities from Annette Island to the Aleutians are thinking about turning on the tap."

Global H2O, for example, a Canadian-based company, has signed a 30-year agreement with the town of Sitka to export 18.2 billion US gallons (about 69 billion liters) per year of glacier water to China, where it is to be bottled in one of that country's "free trade zones," which make extensive use of cheap labor. In order to transport the bulk water from Sitka to markets in China and elsewhere, Global has formed a "strategic alliance" with the Signet Shipping Group, a U.S. company based in Houston, Texas, which has a fleet of supertankers. Each Signet supertanker (50,000 deadweight tons) is expected to carry over 330 million liters (about 87 million US gallons). Global also has a contract with Singapore. "But to supply Singapore on a regular basis," said Global's CEO Fred Paley in June 1998, "we are looking at converting single-hull supertankers which the oil industry will be decommissioning."

At the same time, the man who was the architect of Alaska's pathbreaking policy on water exports has now followed the general market trend by setting up his own water-exporting company. As Alaska's Director of Water,

Ric Davidge was responsible for initiating the marketing of the state's water and established the policy framework that allowed the export of water. Prior to this he served in the U.S. Department of the Interior as chairman of the Federal Land Policy Group and was a key advisor to both federal and state governments in the cleanup operation of the Exxon Valdez oil spill. He also served in President Reagan's sub-Cabinet as Deputy Assistant Secretary for Fish, Wildlife and Parks. After leaving government for the private sector, Davidge went on to establish his own company, called Alaska Water Exports. In 1999, he formed a consortium of companies, World Water S.A., which includes Japan's NYK Line (Nippon Yusen Kaisha), the world's largest shipping company, operating over seven hundred vessels, including a fleet of supertankers.

Yet both Global H2O and World Water are currently prevented from shipping Alaskan glacier water to California, Arizona, and other thirsty regions of the U.S. because of the *Jones Act*. In the United States, shipment of goods from one U.S. port to another is subject to a *Jones Act* provision which insists on the use of U.S. vessels crewed by U.S. sailors. At this point, neither Global nor World Water are able to fully meet these conditions. So while they can use supertankers to ship Alaskan fresh water supplies to China and the Middle East, they are not allowed to transport bulk water to Los Angeles or San Diego at the present time.

Alaska, however, is not the only North American supply base for fresh water supertanker shipments around the world. In the spring of 1998, the Ontario government approved a plan proposed by Nova Group, a Canadian export company, to ship millions of liters of Lake Superior water by tanker to Asia. The approval, however, was rescinded after an outcry from the International Joint Commission and people living around the Great Lakes on both sides of the Canada-U.S. border. Then U.S. Secretary of State Madeleine Albright also issued an official complaint, declaring that the United States had shared jurisdiction over Lake Superior. Elsewhere in Canada, a company called the McCurdy Group has applied for the right to export 52 billion liters (about 13.7 billion US gallons) a year from pristine Gisborne Lake, located in a wilderness area of Newfoundland. The company hopes to ship the water to markets in the Middle East by supertanker.

GRAND CANALS

In recent years, there has been a revival of more traditional methods of transporting bulk water — namely, canal schemes. Given new engineering and construction technologies, canals are now being planned and developed on a cross-continental basis. Suez, the global water giant, for example, is planning to build another Suez-type canal, this time in Europe. Recently, the corporation announced its intention to construct a 160-mile artery to transport water from the Rhône River through France to the Catalonian capital, Barcelona.

Although numerous canal schemes for bulk water transfers are being planned and developed all over the world, nowhere have the canal dreams been as grandiose as in North America. Here, a number of grand canal schemes have been devised to reroute natural river systems in order to deliver huge supplies of water from Canada to the U.S. One of the more prominent of these proposed schemes was literally named the GRAND Canal — the Great Recycling and Northern Development Canal. As originally conceived, the GRAND Canal plans called for the building of a dike across James Bay at the mouth of Hudson Bay (both of which now flow north) in northern Quebec, thereby creating a giant fresh water reservoir of about eighty thousand square kilometers (about thirty thousand square miles) from James Bay and the 20 rivers flowing into it. Through a system of dikes, canals, dams, power plants, and locks, the water would then be diverted from the reservoir and rerouted southward down a 167-mile (269-kilometer) canal, at a rate of 62,000 imperial gallons (about 282,000 liters) a second, into Georgian Bay in Lake Superior. From there, the water would be flushed through the Great Lakes into canals for delivery into markets in the American Midwest and the U.S. Sun Belt.

The GRAND canal vision was widely promoted in the mid-1980s by Quebec Premier Robert Bourassa, who presided over the construction of the massive James Bay hydro project, and by Simon Reisman, Canada's chief negotiator for the U.S.-Canada Free Trade Agreement. Before being appointed Canada's free trade negotiator, Reisman had been an Ottawa lobbyist for Grandco, a four-company consortium promoting the proposed Can$100 billion GRAND Canal. The Grandco consortium was headed by Thomas Kierans, the engineer who had initially developed the scheme and was one of Canada's leading managers of investment capital. One of the major partners in the consortium was the Bechtel Corporation, the U.S. engineering and construction giant that has become, more recently, a key player in water privatization. As trade negotiator, Reisman initially used the GRAND Canal idea as a means to interest the U.S. in free trade, declaring: "In my judgement, water is the most critical area of Canada-U.S. relations over the next 100 years . . . How quickly this issue develops and how much attention is paid depends on how critical the American water shortage is."

Another mega-canal proposal was the NAWAPA — the North American Water and Power Alliance — designed to carry bulk water from Alaska and northern British Columbia for delivery to 35 U.S. states. By building a series of large dams, water from the Yukon, Peace, and Liard rivers would be trapped in the Rocky Mountain trench, a giant reservoir about eight hundred kilometers (about five hundred miles) long, flooding approximately one-tenth of British Columbia. Through this reservoir, a canal would, in effect, be created from Alaska to Washington state, where it would be rerouted to supply water through existing canals and pipelines to customers in the 35 states. The annual volume of water to be diverted through the NAWPA canal scheme would be roughly equivalent to the average total yearly discharge of the St. Lawrence River system.

Initially planned by a group of California entrepreneurs, the NAWAPA canal would have cost an estimated one-half trillion US dollars to build. Although both the GRAND and the NAWAPA canal schemes have been put on the back burner, largely because their huge costs made them financially nonviable when they were first conceived, there are signs that they may soon be revived. As the *Canadian Banker* magazine put it in 1991: "The concept of NAWAPA . . . remains a potentially awesome catalyst for economic and environmental change." In market terms, it is the extent of U.S. demands for water that will determine whether these canal schemes are financially viable or not in the future. Yet the potential ecological costs of these megaprojects are staggering. As Marq de Villiers writes in his book *Water*, the proposed NAWAPA itself "would do as much damage to the environment as all the river diversions combined in America."

Elsewhere, numerous other grand canal schemes continue to be developed. The gargantuan Three Gorges Dam project in China includes a plan to divert water from the mighty Yangtze River to Beijing, primarily to serve industrial and commercial interests. Ten thousand workers have almost finished drilling a 420-kilometer (260-mile) network of tunnels designed to drain water in the middle stretch of the Yangtze, from where it will be channeled through either a high mountain range or a new 1,230-kilometer (764-mile) canal to Beijing. This project, says the Worldwatch Institute, would be like altering the course of the Mississippi River in the U.S. to service the water needs of Washington, D.C. Moreover, for different strategic objectives, there appear to be plans to build more Panama-type canals across Central America. According to Andréas Berreda, a cartographer and professor of geography at the University of Mexico, there are as many as five projects now on the drawing boards to build canals through Central America, mainly designed to greatly expand the multibillion-dollar shipping industry between Europe and China as part of the new worldwide trade system. While these canals have not been planned for bulk water exports, they could well be used for increasing supertanker shipments of water.

WATER BAG SCHEMES

One of the new technologies rivaling supertankers as a means of transoceanic bulk fresh water transport is the manufacture of huge sealed bags hauled by tugboats. According to Medusa, a Canadian-based company specializing in research and development of this technology, it is possible to produce a water bag with the carrying capacity of five supertankers at about 1.25 percent of the cost. If this new water-bag technology was proven to be efficient, supertankers carrying 400,000 cubic meters (about 14 million cubic feet) of water would no longer be economical. The objective, says Medusa, is to produce water bags that have a much greater capacity than supertankers — that is, between 500,000 cubic meters and three million cubic meters (between about 17 million and 106 million cubic feet). (Each cubic meter contains 240 US gallons, or 908 liters, of water.) What's more, the water bags can be hauled by conventional tugs or the type of vessel used for offshore oil rigs, with minor modifications.

Medusa also maintains that this technology can be developed to meet specific requirements. The size and shape of the water bags can be flexibly designed to suit diverse situations, taking into consideration such factors as the fabric used, towing costs, annual delivery volumes, and the coastal characteristics of the delivery route itself. A water-bag unit with a carrying capacity of 1.75 million cubic meters (about 62 million cubic feet) could be designed, says Medusa, in a streamlined shape with a flat top and bottom, 650 meters long by 150 meters wide and 22 meters deep (about 2,100 by 490 by 72 feet). But since a water bag of such large dimensions would require more research and testing before it was ready for production, Medusa decided to concentrate its work in the year 2000 on the production of smaller units with carrying capacities of about 100,000 cubic meters (about 3.5 million cubic feet).

Meanwhile, several corporations have already begun specializing in the use of this new technology for bulk water exports. In the U.K., the Aquarius Water Transportation Co. (whose corporate investors include Suez) began the first commercialized deliveries of fresh water using polyurethane bags towed by tugboats. The company's bag fleet consists of eight 720-ton and two 2,000-ton water-capacity bags (which hold 2 million liters, or about half a million US gallons). Aquarius, whose corporate investors include the water service conglomerate Suez, has been delivering water to the Greek Islands since 1997 using water-bag technology. The polyurethane bags are manufactured in the United Kingdom, where they are tested and approved by an independent government agency. In addition to the bags used to ship water to the Mediterranean, Aquarius uses the larger, 2-million-liter bags for short-haul deliveries. While bags ten times larger have been designed, more capital investment is required to produce them, so Aquarius has not purchased any of that size. However, the company predicts that its market for water-bag shipments will soon exceed 200 million metric tons a year, and the company plans to secure contracts in other Mediterranean islands, Israel, and the Bahamas.

In Norway, the Nordic Water Supply Co. has developed a sea-water and UV-resistant water bag made of polyester fabric coated on both sides with a polymer mixture. Since 2000, Nordic has been using this kind of bag, towed by tugboat, to transport fresh drinking water from the Turkish port of Antalya to northern Cyprus. Larger than the bags Aquarius uses, Nordic's 160-meter-long bag carries 5 million US gallons (19 million liters) of fresh water. Tested for year-round use, the bag is designed to weather stormy North Sea conditions. In December 2000, however, Nordic lost one of the bags it was transporting about five miles off the coast of Cyprus. The company recovered from the incident, and in 2001, Nordic went on to develop contracts for bulk water-bag shipments in Greece, the Middle East, Madeira, and the Caribbean.

A California-based entrepreneur, Terry Spragg, has pioneered another method of water-bag delivery. Convinced that it makes more economic sense to transport large volumes of water under one tow, instead of smaller volumes under several tows, Spragg has been developing a train method of delivery whereby up to 50 smaller bags (holding approximately 17,000 cubic meters each, or about 600,000 cubic feet) are towed. Assisted by engineering specialists at the Massachusetts Institute of Technology and the CH2M-Hill Co., Spragg has designed what *Water Resources* magazine describes as "a unique high strength zipper system that will connect the bags via a non-watertight fabric sleeve that can fill with seawater and gradually express it to ease stresses posed by differential movement between bags under tow." While working out the numerous technical problems associated with hauling a train of water bags at sea, Spragg's contracts have been focused on transporting water from northern regions to southern California.

At present, water-bag technology is still in its early stages of development, and no one can be sure that it will prove to be economically and ecologically viable. Although governments like Turkey have expressed strong interest in water-bag delivery, more capital investment is needed before the technology can be developed sufficiently to become a suitable replacement for supertankers. As a mode of bulk water delivery, water bags are a great deal cleaner and safer than supertankers, but that does not necessarily make them ecologically sound. As long as the fresh water itself is extracted from its natural location, there will be negative environmental repercussions.

BOTTLED WATER

The one water-export method that has been taking off is bottled water. It is among the fastest-growing and least-regulated industries in the world. In the 1970s, the annual volume of water bottled and traded around the world was 300 million US gallons (about 1 billion liters). By 1980, the figure had climbed to 650 million US gallons (about 2.5 billion liters), and toward the end of the decade, 2 billion US gallons (7.5 billion liters) of bottled water were being consumed in countries around the world. But in the past five years, the volume of bottled water sales has skyrocketed, and in 2000, 22.3 billion US gallons (84 billion liters) of water were bottled and sold. Moreover, one-quarter of all the water bottled was traded and consumed outside its country of origin.

Among the brand name products are Perrier, Evian, Naya, Poland Spring, Clearly Canadian, La Croix, Purely Alaskan, and many more. Nestlé is the world market leader in bottled water, with no fewer than 68 brands, including Perrier, Vittel, and San Pellegrino. As a past chairman of Perrier put it: "It struck me . . . that all you had to do is take the water out of the ground and then sell it for more than the price of wine, milk, or for that matter, oil." While bottled water may have started out as a pampered Western consumer affectation, Nestlé has found a growing market niche for bottled water in nonindustrialized countries where safe tap water is rare or nonexistent. In these countries, its main product line is Nestlé Pure Life, a low-cost purified tap water with added minerals. Marketed on a platform of "basic wholesomeness," Nestlé Pure Life has sold well in Pakistan and Brazil, as have some of the corporation's other bottled water products in China, Vietnam, Thailand, and Mexico.

In 2000, worldwide sales in bottled water were estimated to be around US$22 billion. However, this figure pales in comparison to the 1998 figures presented by the Euromonitor statistical agency, which indicate there were global sales amounting to US$36 billion in 53 countries in that year. Regardless of which figures are used as the benchmark, the bottled water industry has been growing at an astounding rate. Besides Nestlé, other giants of the global food and beverage industry have also become purveyors of bottled water, including Coca-Cola, PepsiCo, Procter & Gamble, and Danone. With the entry of the big soft drink giants, market growth is expected to accelerate further. PepsiCo is currently leading the way with its Aquafina line, while Coca-Cola has launched a North American line under the name Dasani at the same time as continuing to market its international label, Bon Aqua.

Yet in contrast to the market image of "pure spring water" that is projected by the industry, bottled water is not always safer than tap water and some is less so. That was the conclusion of a March 1999 study by the U.S.-based Natural Resources Defense Council (NRDC) which found that one-third of the 103 brands of bottled water it studied contained levels of contamination, including traces of arsenic and *E. coli*. One-quarter of all bottled water is actually taken from the tap, though it is further processed and purified to some degree, said the NRDC study, and in many countries, bottled water itself is subject to less rigorous testing and lower purity standards than tap water. "One brand of 'spring water,'" reported the NRDC, " . . . actually came from a well in an industrial facility's parking lot, near a hazardous waste dump. And periodically was contaminated with industrial chemicals at levels above FDA standards."

In addition, the marketing hype about bottled water being more environmentally friendly and healthier than tap water is also misleading. In terms of nutritional value, according to the United Nations Food and Agricultural Organization (FAO), bottled water is no better than tap water. The idea that bottled "spring" or "natural" water contains near-magical qualities and great nutritive value is "false," declares a 1997 FAO study on "Human Nutrition in the Developing World." "Bottled water may contain small amounts of minerals such as calcium, magnesium, and fluoride, but so does tap-water from many municipal water supplies." The FAO report also cites a study "comparing popular brands of bottled water [which] showed that they were in no way superior to New York tap-water." And as far as environmental responsibility is concerned, a study released by the World Wildlife Federation (WWF) in May 2001 shows that the bottled water industry uses 1.5 million tons of plastic every year, and when plastic bottles are being manufactured or disposed of, they release toxic chemicals into the atmosphere. Furthermore, since a quarter of all bottled water produced is for export markets and transportation fuel results in carbon dioxide emissions, the WWF report contends that the transportation of bottled water is a contributing factor to the problem of global warming.

Worse still, the relentless search for secure water supplies to feed the insatiable appetites of the water-bottling corporations is having damaging effects. In rural communities throughout much of the world, the industry has been buying up farmland to access wells and then moving on when the wells are depleted. In Uruguay and other parts of Latin America, foreign-based water corporations have been buying up vast wilderness tracts and even whole water systems to hold for future development. In some cases, these companies end up draining the water system of the entire area, not just the water on their land tracts.

What's more, the bottled water corporations generally pay no fee for the water they remove according to so-called private property rights — although water is a part of the commons. In Canada, for example, where the amount of water extracted by the bottling industry has grown by 50 percent in the past decade, bottlers have a legal right to take about 30 billion liters (about 8 billion US gallons) a year — approximately 1,000 liters (or about 265 US gallons) for every person in the country. Close to half of all this bottled water is exported to the United States. Yet unlike the oil industry, which pays royalties, and the timber industry, which pays stumpage fees to the government, the water-bottling business is not required to pay fees for the extraction of water in most Canadian jurisdictions. According to a survey conducted by the *Globe and Mail* newspaper, every provincial government within Canada, with the exception of British Columbia, allows these companies to extract water from lakes, rivers, and streams without charging them a fee. And the fees collected by the B.C. government every year for water extraction in that province amount to a minuscule Can$25,000.

The global gaps between rich and poor are also dramatically mirrored in the marketing strategies of the bottled water corporations. In its 1999 study, the U.S.-based Natural Resources Defense Council (NRDC) reported that some people pay up to ten thousand times more per gallon of bottled water than they do for tap water in their communities. For the same price as one bottle of this "boutique" consumer item, says the American Water Works Association, one thousand gallons of tap water could be delivered to a person's home. Ironically, the same industry that contributes to the destruction of public water sources — in order to provide "pure" water to the world's elite in plastic bottles — is peddling its product as being environmentally friendly and part of a healthy lifestyle.

COKE WATER

The well-known "Cola Wars" between PepsiCo and Coca-Cola are just beginning to play out in the theater of the bottled-water trade. According to the annual Global Fortune 500 listings, the two soft drink giants ran neck and neck in 2000, ranked 233 and 234, with Coca-Cola's US$20.458 billion in total revenues nudging out PepsiCo's US$20.438 billion. Water, of course, has always been an essential ingredient in the manufacturing of soft drinks, and both PepsiCo and Coca-Cola have had to secure access to supplies of clean water for their production. But now both soft drink giants have become major players in the bottled water market, PepsiCo with its Aquafina product and Coke with Dasani. Since the bottled water industry could end up leading the way in setting the worldwide price for water, it is important for us to take a closer look at the emerging cola water wars, particularly Coke's challenge to Pepsi.

Pepsi got the head start in bottled water by putting Aquafina on the market in 1994. Although Coke owned bottled water brands such as Bon Aqua in 35 other countries, it was not until five years later, in 1999, that Coke decided to challenge Pepsi on this front by launching its Dasani product for distribution in North America. In the year 2000, Pepsi's Aquafina led the way in global bottled water sales with a 7.8 percent market share, while Coke's Dasani was ranked fifth, with 4.9 percent. During the first quarter of 2001, however, Pepsi and Coke were ranked first and second in the critically important U.S. market. Here, Pepsi's Aquafina captured 15.1 percent of the U.S. market, up 59.4 percent, while Coke's Dasani recorded an even greater increase of 123.9 percent, for 8.7 percent of U.S. customers. What these trend lines indicate, say several market analysts, is that dozens of the smaller brands are beginning to fall off the chart as the two soft drink giants gradually emerge as the dominant players in the bottled water industry, expanding their market share every year through their vast distribution networks around the world.

Unlike most other water bottlers, however, Pepsi and Coke specialize in "purified water," rather than "spring water." Moreover, Aquafina and Dasani are sold as purified local water, since they come out of the tap from municipal water systems. Instead of drawing water from the ground and transporting it long distances from natural springs, Pepsi and Coke run municipal water through a "reverse osmosis" filter system, add some minerals, and peddle the result as purified water. Both soft drink giants are in a position to use this procedure on a large scale because they have local bottling companies all over the world. The water they use from municipal systems generally costs them only a fraction of a cent per liter, and after they purify and bottle it, they sell it at approximately one dollar per liter. Although the filtering systems used by Pepsi and Coke take more impurities out of the water than municipal water plants do, observers point out that purified bottled water is sometimes no safer than what comes out of the faucet in most North American communities.

Coke's entry into the bottled water market came after a prolonged internal debate. Roberto Goizueta, Coke's CEO from 1981 to 1997, proclaimed in 1986 that by the early 21st century, people all over the world would be following the pattern of Americans who consumed soft drinks, including Coke, more than any other liquid. In other words, Coke was betting that soft drinks would beat out ordinary tap water as the number one way that humanity would get hydrated in the future. But during the late 1990s, the soft drink market flattened out, while bottled water, popular because of its convenience, took off — especially in industrialized countries. For Coke, the issue was how to make bottled water profitable. After all, Coca-Cola had made its fortune in soft drinks by selling concentrate, or syrup, to independent bottling franchises, which then added water and carbonation before distributing the product. With bottled water, Coke couldn't sell any syrup. However, purified water would require minerals and small amounts of potassium and magnesium, to make the water taste better. So instead of selling concentrate, Coke decided to sell its bottlers mineral packets. This way, it could use the same system that had been such a financial success in the production of Coke.

In their marketing strategies, both Pepsi and Coke are counting on "brand loyalty" to carry them to the top of the bottled water industry, which they eventually want to dominate. To sell their purified water product in North America and Europe, both soft drink giants are focusing on issues of healthy lifestyles. Pepsi's advertising campaigns note that since the human body is 70 percent water, "every part of your body needs pure water" and people can replenish this vital liquid by drinking Aquafina. Similarly, Coke's "Life Simplified" campaign promotes Dasani as replenishing the body's water, thereby restoring balance through relaxation, health, and wellness in the midst of busy and stress-filled lives. And the Dasani "Treat yourself well. Everyday" campaign is specifically targeted at women between the ages of 25 and 49. In May 2001, for example, Coke announced a partnership with iVillage, an online website whose core audience is women, to promote tips from Dasani's "Wellness Team" on cooking, nutrition, fitness, and stress management. "The Healing Garden," a gift box of lavender lotions, bath crystals, and incense sticks, is also provided by the Dasani campaign, along with a "Personal Balance Index." And to capture the teen market, Dasani is now vigorously promoted and sold in every North American school that has signed a marketing contract with Coca-Cola.

Yet Coke also realizes that its greatest potential market lies outside of North America. More than three-quarters of Coke's global revenue comes from international sales. According to Coca-Cola, approximately 17 billion cases of Coke drinks are continually available on the market in two hundred countries around the world. In the first quarter of 2001, Coke's soft drink sales grew ten times faster in Asia and Africa than they did in North America. This is not too surprising, since marketing studies have shown that the red and white Coca-Cola label is recognized by up to 98 percent of teenagers around the world.

Given their worldwide presence so far, Coke is expanding operations in several countries. In July 2001, the business press reported that Coke was planning to use its Latin American bottling operations to expand into what it sees as a considerable market on that continent. In Mexico, which J.P. Morgan investment analysts report as being second only to Italy in per capita consumption of bottled water, Coke has a network of 17 bottling companies, compared to 6 for Pepsi. In Brazil, where Coke has 19 bottling companies and has been selling its Bon Aqua brand of bottled water since

1997, the company plans to increase its market share of sales in purified water. The same goes for Chile, where Coke already has 30.8 percent of the mineral waters market, as well as 69 percent of the soft drink market.

Coke's marketing of both its traditional soft drink and its bottled water brands is based on the human need for hydration. As the company's 2000 annual report says, "We're redefining how consumers get hydrated." According to conventional wisdom, people generally need to drink eight glasses of water a day to rehydrate themselves, but soft drink sodas are significantly less hydrating than plain water. Nutrition experts like Marion Nestle at New York University note that soft drinks not only have a *diminishing* hydration effect, but "a thirsty person would be far more wired than hydrated by eight or more caffeinated sodas a day." In addition, soft drinks are hardly nutritious, contributing to both tooth decay and obesity. " . . . In places where vast majorities lack nutritious calories and clean water," writes Sonia Shah, a former editor/publisher at South End Press in Boston, "it seems a double cruelty to siphon off . . . clean water . . . and adulterate it with brown syrup."

Recently, Coca-Cola has begun contemplating an even more radical way of widening its distribution network. In March 2001, the current CEO of Coca-Cola, Douglas Daft, announced that the company's innovation unit had developed a prototype for bringing Coke right into people's homes through a tap on the kitchen sink. "You would have water mixing automatically with the concentrate," said Daft, " . . . then connect it all up so that when you turn on your tap, you have Coke at home." To remove regional differences in flavor, water pumped through the tap would be purified, then carbonated and mixed with syrup right at the kitchen sink. "There's a lot more to it than that to ensure quality," added Daft, "and it has to be a sealed unit so people can't alter the formula to destroy the value of the brand." In effect, the "home Coke-on-tap" system would fulfill former CEO Roberto Goizueta's desire that the "C" designating the cold tap would come to stand for Coke. It may sound like a pipe dream now, but it's consistent with Coke's long-term goal of redefining how people "get hydrated" and with Goizueta's dream that people all over the world would soon be consuming more soft drinks than water.

Meanwhile, Coca-Cola's own performance record in terms of product quality and promotion should not be forgotten. In Belgium and France, over two hundred people suddenly fell ill in June 1999 after drinking Coke that had become contaminated by the use of a substandard carbon dioxide in an Antwerp bottling plant and by chemical preservatives on the wooden crates used to ship cans from Dunkerque, France. Coca-Cola Enterprises, the licensed bottler for Belgium and most of France, recalled and destroyed some 17 million unit cases of the product from European supermarkets and vending machines. Soon afterwards, bottles of Coke's Bon Aqua brand of mineral water distributed in Poland were found to contain mold and bacterial contamination and had to be recalled.

In the United States, on Earth Day 2000, the GrassRoots Recycling Network named Coke "the champion beverage waste-maker" because, since 1995, it had produced more than 21 billion plastic bottles that had been "wasted, as litter on the streets, in parks and on beaches, or sent to landfills and incinerators." The calculations were made by Dr. Bill Sheeham, Coordinator of the GrassRoots Recycling Network. Based on this group's work, local government officials in Florida, Minnesota, and California have passed resolutions calling on Coke to make new plastic soda bottles with recycled plastic. And among numerous problems reported from Third World countries, a 1991 study by the federal Food Department in Rio de Janeiro showed that poor children aged 6 to 14 suffering from serious malnutrition and protein deficiency had consumed large quantities of Coke since infancy. As one UN official put it more recently, "Instead of breast milk, . . . [Third World] children get Coca-Cola."

In 1980, a worldwide boycott against Coke was ignited when rumors circulated that the company may have had connections with death squads in Guatemala. The concerns were increased when two union leaders were murdered. The allegations against Coke were neither made nor proven in court.

Two decades later, on July 23, 2001, Coca-Cola was taken to court by labor leaders who claimed that Coke should bear some responsibility for what it said was a failure to prevent its Colombian bottlers from bringing in right-wing paramilitaries who allegedly used murder and torture to break up unions at their plants. The case is being brought against Coca-Cola and the bottlers themselves under the *Alien Tort Claims Act*, which allows foreigners to sue U.S. companies for damages caused abroad. Coca-Cola has firmly denied the charges, and the bottlers, according to one news report, had made no comment as of July 29, 2001.

In Atlanta, Georgia, where Coca-Cola has its headquarters, journalists dubbed January 26, 2001, the "Day of the Long Knives," after 21 percent of the company's twenty-nine thousand workers worldwide suddenly lost their jobs so the company could save US$300 million annually. According to the IUF, a food and hospitality services workers' union, Coke's success has been built on a single overriding corporate strategy: "produce, promote and market Coca-Cola brand world-wide while minimizing the number of workers directly employed by the company." By subcontracting out to franchise operators and "anchor bottlers," Coke has been able to avoid taking on the vast numbers of employees that would otherwise be required. While there is nothing illegal about operating a transnational corporation through franchises, this action allows Coca-Cola to have fewer direct employees to whom it is responsible.

In the U.S., Coca-Cola has also been charged with, and convicted of, racial bias. In 1999, eight black former employees sued the company, accusing it of denying fair pay, promotion, raises, and performance reviews to blacks. On November 16, 2000, Coca-Cola was ordered by the court to pay approximately US$190 million to some two thousand black workers.

GLOBAL CARTEL

The question still remains as to whether a global cartel will emerge to control water exports by the year 2010. If the OPEC model were followed, the countries containing large supplies of fresh water in the form of lakes, rivers, and glaciers would constitute such a cartel. Studies by the highly respected Russian hydrologist Igor Shiklomanov, described by Peter Gleick in his book *Water in Crisis*, identify the countries with the most fresh water in the world. Twenty-eight of the world's largest fresh water lakes, he writes, account for 85 percent of the volume of all lake water, including Russia's Lake Baikal, Africa's Lake Tanganyika, and Lake Superior on the U.S.-Canadian border. As the world's largest lake system, the Great Lakes together account for 27 percent of global lake volumes. The world's largest

25 rivers include: 11 in Asia (the Ganges, Yangtze, Yenisei, Lena, Mekong, Irrawaddy, Ob, Chutsyan, Amur, Indus, and Salween); 5 in North America (the Mississippi, St. Lawrence, Mackenzie, Columbia, and Yukon); 4 in Latin America (the Amazon, Paraná, Orinoco, and Magdalena); 3 in Africa (the Congo, Niger, and Nile); and 2 in all of Europe (the Danube and the Volga).

Based on these statistics, Brazil contains the highest proportion of fresh water resources (approximately 20 percent of all global supplies), followed by the former Soviet Union countries at 10.6 percent, China at 5.7 percent, and Canada with 5.6 percent. These figures, however, do not include the vast potential sources of glacial water found in the Arctic, Alaska, Greenland, Siberia, and the Antarctic, or in mountain ranges like the Alps. When glacial water sources are included, countries like Norway, Austria, and the United States (Alaska) can be added. But given this cross-national picture, it is not at all clear that an OPEC-style water cartel will be formed. With the exception of the longstanding American interest in obtaining water from Canada, none of these nation-states has yet demonstrated either the interest or the ability to organize such a cartel. In fact, it is more likely that transnational corporations rather than national governments will seize control of fresh water supplies around the world and that they will set the stage for a future global water cartel.

Yet for the moment, the only corporate players that have focused their energies on seizing control of untapped fresh water supplies for bulk water exports are small, independent companies working in consortium with larger corporations. Global H2O, as we have seen, has formed an alliance with Signet Shipping for long-distance transport of Alaskan glacial water to China and other markets.

Similarly, as mentioned earlier, Alaska Water Exports' Rick Davidge, known as "Alaska's Water Czar," has formed a consortium, World Water S.A., which has secured access to glacial water supplies in both Alaska and Norway. Its partners now include the NYK Line of Japan for supertanker transport and Nordic Water Supply of Norway for water-bag transport. Global H2O and World Water may be prototypes of the kind of corporate joint ventures that will dominate global water exports in the future, though this industry is only in its early stages.

Meanwhile, major corporate players in other industries, with greater access to sources of capital investment, could also become movers and shakers in bulk water exports. As the market value of water continues to rise, major players in the energy industry like Exxon, Shell, and British Petroleum, as well as energy service corporations like Enron, could decide it is in their interest to secure control over bulk water supplies for transport by pipeline or supertanker. Similarly, if demand increases for cross-continental transfers of bulk water through grand canals, corporations that already combine engineering, construction, and water interests like Suez, Bechtel, and RWE will likely become key players in water exports. More than likely, these are the kinds of water ventures that George Soros' Global Power Fund would finance, channeling investment capital from the World Bank and GE Capital (the largest private source of investment capital in the world, owned by General Electric). And regardless of what happens on these water export fronts, the bottled water industry will continue to play a key role in establishing world water prices, led by the soft drink giants, PepsiCo and Coca-Cola.

For the time being, however, it looks as if the concentration of power in the water export industry will take forms different than these — along both sectoral and regional lines — depending on the extent to which the world experiences water shortages and crises. Over the next five years or so, the major corporate players in various industries will no doubt determine for themselves, based on market demands and profitable opportunities, whether bulk water can be most efficiently transported by pipelines or canals, supertankers or water bags. At the same time, strategic links between bulk water supplies and market demands are likely to become more focused on a regional basis, with Norway and Austria's supplies targeted for water-shortage areas in Europe and the Middle East, Brazil's water going to other parts of Latin America, and Canadian/Alaskan water being exported to thirsty regions of the United States and Mexico. As the emerging global water export industry takes shape along these lines in the coming years, there will undoubtedly be closer and closer ties between corporations and governments. After all, corporations still need governments to provide them with the sense of political, and perhaps moral, legitimacy they need to control and sell such an essential, life-giving resource on a for-profit basis.

Yet, this sense of legitimacy may soon be increased. As this book was going to press, there were reports that at least two dozen countries were on the verge of taking initiatives to allow water exports.

The more this corporate-government nexus solidifies around water exports, the more we can expect institutions of global economic governance like the World Trade Organization, the International Monetary Fund, and the World Bank to play a decisive role in determining what kind of world water cartel is likely to be in place by the year 2010.

GLOBAL NEXUS

*How international trade and financial
institutions have become the tools of
the transnational water companies*

Early one morning in April 2000, a five-foot, slightly built, 45-year-old machinist from Cochabamba, Bolivia, by the name of Oscar Olivera, climbed aboard a plane, and for the first time in his life, left his homeland. He was flying to Washington, D.C., where he hoped to confront the Director of the World Bank, James Wolfensohn, with a message from his people.

In 1998, the World Bank notified the Bolivian government that it would refuse to guarantee a US$25 million loan to refinance water services in the city of Cochabamba unless the local government sold its public water utility to the private sector and passed on the costs to consumers. In response, the Bolivian authorities saw to it that control over the Cochabamba water utility was handed over to Aguas del Tunari, a newly formed subsidiary of the U.S. construction and water giant Bechtel. The World Bank then granted monopolies to private water concessionaires, called for full-cost water pricing, pegged the cost of water to the U.S. dollar, and instructed the Bolivian government that the loan monies could not be used to subsidize the poor for water services.

After seeing their water rates jacked up by nearly 35 percent, the people of Cochabamba took to the streets by the tens of thousands, shutting down their city for four straight days through a series of strikes and block-ades in January 2000. The protest was coordinated by the Coordiadora de Defense de Agua y la Vida (the Coalition in Defense of Water and Life) led by Oscar Olivera. Public opinion polls showed that 90 percent of Cochabamba's citizens wanted Bechtel's subsidiary to return the city's water system to public control. After a week of escalating protests, Bolivian President Hugo Banzer placed the country under martial law and announced that the government would break its contract with Bechtel. But not before a 17-year-old boy was shot to death.

When asked to comment directly on the Bolivian protests, World Bank Director Wolfensohn maintained that giving public services away to people leads inevitably to waste and that countries like Bolivia need to have "a proper system of charging" for water. He was echoing a World Bank report from June 1999, which declared: "No subsidies should be given to ameliorate the increase in water tariffs in Cochabamba," arguing that all water users, including the very poor, should be called to bear the full cost of the water system and its proposed expansion. Yet Wolfensohn flatly denied that the privatization scheme was directed against the poor.

"I'd like to meet with Mr. Wolfensohn to educate him on how privati-zation has been a direct attack on Bolivia's poor," replied protest leader Olivera. "Families with monthly incomes of around $100 have seen their water bills jump to $20 per month — more than they spend on food. I'd like to invite Mr. Wolfensohn to come to Cochabamba and see the reality he apparently can't see from his office in Washington, D.C."

⌒

By the time Oscar Olivera landed in Washington, D.C., he found himself in a surprisingly familiar atmosphere. Tens of thousands of people from all over the United States, along with representatives of social movements around the world, had gathered together on the streets of the nation's capital to protest the policies and programs of the World Bank and the

International Monetary Fund (IMF). Since the early 1980s, the World Bank and the IMF had been imposing "Structural Adjustment Programs" on Third World countries as conditions for renewing their financing and international debt payments. Through these programs the governments of Third World countries were compelled to undertake a series of radical measures, ranging from the sell-off of public enterprises to pay back debt loans to massive reductions in public spending on health, education, and social services. These structural changes, in turn, have had devastating impacts on the living conditions of the poor majority in these countries over the past decade and a half. In recent years, one of the prime conditions for the renewal of World Bank and IMF loans has been the privatization of a country's public water and sanitation facilities.

This process has had repercussions on the daily lives of thousands of people, especially poor people, in Third World countries. That's why the struggle in Cochabamba, led by Olivera and La Coordiadora, brought a human face to the protests in Washington, where the World Bank and the IMF were holding their annual meetings in April 2000. As the story of Cochabamba reveals, the World Bank's demands are designed primarily to benefit global water corporations like Bechtel. It also shows how much the major corporate players in the water industry depend on these international financial institutions to build a worldwide water market. But the power and influence of the global water industry does not end there. The close ties between corporations and governments have created a network of institutions for global economic governance, which has established a body of rules for finance, trade, and investment that can now be effectively used by both water service companies and water exporters.

CORPORATE MASSAGES

The big corporate players in the global water industry have not left things to chance, since they know all too well that they must prime the pump for both privatization and exports of water in the global economy. The primary institutions of global economic governance — the World Trade Organization, the World Bank, and the IMF — are indispensable in providing the financial and legal leverage required to build a global water market. Getting governments in key countries onside is also a major factor in implementing the global water industry's agenda. As Rebecca Mark, the former CEO of Azurix, put it, "We're going to be in their face." To carry out these strategies, the major corporate players realized they needed to develop mechanisms for political massaging. As a result, an integrated network of lobby organizations, professional associations, and corresponding political machinery has been put in place.

In 1992, two events laid the foundation stones for the formation of a network of international water agencies: the International Conference on Water and the Environment (ICWE) in Dublin and the United Nations Conference on Environment and Development (UNCED) in Rio de Janeiro. In particular, three interrelated agencies emerged: the Global Water Partnership (GWP), the World Water Council (WWC), and the World Commission on Water (WCW) for the 21st Century. On the surface, each of these international water agencies appears to be neutral because in theory, they exist to facilitate dialogue between the various stakeholders and to bring about a more sustainable management of water resources. But a closer look reveals that these agencies promote the privatization and export of water resources and services through close links with global water corporations and financial institutions.

The Global Water Partnership was established in 1996 to "support countries in the sustainable management of their water resources." Its operating principle, however, is the recognition that water is "an economic good" and "has an economic value in all its competing uses." This basic principle lies at the core of the GWP's main programs to reform water utility systems and water resource management in countries around the world. The chair of the GWP's Steering Committee is Ishmail Serageldin, Vice-President of the World Bank, and the GWP is funded by both government aid agencies from countries like Canada, Denmark, Finland, Germany, Luxembourg, the Netherlands, Norway, Sweden, Switzerland, and the United Kingdom and international finance agencies like the World Bank, the United Nations Development Programme, and the Ford Foundation.

The World Water Council, also formed in 1996, sees itself as a policy think tank whose main task is to provide decision makers with advice and assistance on global water issues. The WWC's 175 member-groups include leading professional associations, global water corporations, UN organizations, government water ministries, and financial institutions, as well as representatives of some nongovernmental organizations, policy makers, scientists, and the media. Along with the GWP, the WWC played a key role in organizing the second World Water Forum in The Hague in the year 2000, to promote private-public partnerships as the only solution to the global water crisis. The WWC also spearheaded the *World Water Vision* report, in which 85 individuals and groups (many of whom were clearly linked with global water corporations and related bodies) outlined a water privatization agenda.

The third agency that has been created with the stated purpose of fostering sustainable use of water resources is the World Commission on Water for the 21st Century, formed in 1998. Led by the World Bank's Ishmail Serageldin (who is also chair of the GWP's Steering Committee and Vice-President of the World Bank), the Commission is composed of 21 eminent personalities from around the world. Along with the governments of Canada and the Netherlands, the WCW is officially supported by all the major UN agencies with a mandate related to water — the UN Educational, Scientific and Cultural Organization (UNESCO), the UN Development Programme (UNDP), the UN Food and Agricultural Organization (FAO), the UN Environment Programme (UNEP), the World Health Organization (WHO), and UNICEF. Given the Commission's direct ties to the GWP and the WWC, the marketing of water resources and services is likely to be a major factor in its vision for the 21st century.

Representatives of the global water corporations are strategically placed at the top levels of all three of these global agencies. Among them are a number of corporate players from Suez. In 1999, for example, René Coulomb, former Director of Suez, was Vice-President of the World Water Council and an influential member of the Steering Committee of the Global Water Partnership. Ivan Chéret, Senior Advisor to the Chairman of Suez, served on the Technical Advisory Committee of the GWP, and Jérome Monod, chair of Suez's Supervisory Board, was a member of the World Commission on Water. Margaret Catley-Carlson, former President of the Canadian International Development Agency (CIDA) and currently Chair of the Suez-sponsored Water Resources Advisory Committee, is also a member of the World Water Commission.

At the same time, the water corporations have their own network of industry associations for promoting their own projects for water privatization and exports, lobbying governments for legislative and financial assistance, and building community or public support for their agenda. One of these is the International Private Water Association, whose membership includes key players in the industry like Vivendi-U.S. Filter; Enron-Azurix; and Bi-Water, a British-based water service corporation. Organized "to promote global opportunities for private water project development globally" by arranging meetings with government water ministries and local authorities, it has task forces covering Europe, the Middle East, Africa, Asia, North America, and Latin America. Some countries also have their own industry associations. In the United States, the National Association of Water Companies (NAWC) is set up to represent the "private and investor-owned water utility industry." More specifically, the NAWC develops strategies in response to federal or state legislative initiatives and decisions made by regulatory agencies that affect the water industry or markets.

The corporate massage also involves promoting a soft public image. In response to the United Nation's Water Decade (1981–1990), British water corporations, led by Severn Trent, set up WaterAid as a nongovernmental organization (NGO) with this stated goal: "to help poor people in developing countries achieve sustainable improvements in their domestic water supply, sanitation and associated hygiene practices." In practice, WaterAid promotes a public image of concern for the struggles of water-deprived Third World peoples, but meanwhile, private corporations carry on with business as usual. Another association, Business Partners for Development, initiated by the World Bank, contains a "water clusters initiative" to promote "good private sector practice" for the delivery of water and sanitation services to urban poor populations in nonindustrialized countries. As a business association, it works in collaboration with government and certain civil society organizations. And there are also close cross-connections between the different associations and corporations. In

1998–99, for instance, WaterAid, Vivendi, and the World Bank convened a series of international meetings on water issues on behalf of the Business Partners for Development.

In addition to maintaining close working relationships with the World Bank and other global financial bodies, the big water corporations have strategically positioned themselves to play an effective role in the World Trade Organization, especially in the negotiations to establish a new set of global rules for cross-border trade in services. Two powerful lobby machines have been established to advance corporate interests in the WTO service negotiations — namely, the U.S. Coalition of Service Industries and the European Forum on Services. (See Chapter 5.) Water giants Vivendi and Enron are both active members of the U.S. coalition, and Vivendi and Suez are key players in the European Forum on Services. As we saw in Chapter 5, Suez and Vivendi are the world's two largest water service corporations, while Enron is a major multisector service corporation with water interests. And Vivendi is one of only three corporations to be an active member of both lobby machines. Along with the world's largest corporate service providers from other industries, including banking, telecommunications, energy, health care, education, entertainment, postal services, engineering, and social services, Vivendi, Suez, and Enron will effectively be writing a new set of rules that will govern the marketing and trading of services in the global economy. Specifically, they are in a pivotal position to see that these rules promote the privatization and export of water.

International Finance

When it comes to financing water services in the nonindustrialized countries, the main sources of funding are international lending institutions like the International Monetary Fund and the World Bank. The IMF is the multilateral lending vehicle for the central banks of governments, while the World Bank operates mainly as the multilateral lending vehicle for the private banks. The policies and programs of both, however, are closely intertwined. This global financial architecture is further buttressed by a network of regional development banks such as the European Investment

Bank, the Inter-American Development Bank, the Asian Development Bank, the African Development Bank, the European Bank for Reconstruction and Development, and the Islamic Development Bank. In particular, the global water lords have been successful in using not only the World Bank and the IMF, but also most of the regional banks, in their bids to take control of the water systems in many nonindustrialized countries.

With the World Bank, there are two organizational divisions that serve the interests of the global water lords. The first is the International Bank for Reconstruction and Development (IBRD), which provides loans to governments and is in the position to impose conditions like the privatization of public water systems. In 1999, for example, the World Bank forced Mozambique to privatize its water services as a condition attached to a loan that would finance infrastructure development and extend the country's debt relief. Through the IBRD, the World Bank worked in collaboration with the Africa Development Bank and other funding agencies to provide a US$117 million loan to Mozambique. As a condition of the IBRD loan, Mozambique was compelled to privatize its water services. The beneficiary of the long-term contract was Bouygues-SAUR, as SAUR, the water company owned by Bouygues, picked up the long-term contract for water and sanitation services for 2.5 million people, which generates about US$9 million a year in revenues. This pattern has been followed for World Bank loans in many other nonindustrialized countries, including the Cochabamba, Bolivia, case described at the beginning of this chapter, in which a Bechtel subsidiary initially acquired the water concession for that city. The World Bank instructed the Bolivian government that the city of Cochabamba would have to privatize its public water utility as a condition of receiving a US$25 million loan.

At the same time, the World Bank provides capital financing directly to the major water corporations themselves through its International Finance Corporation (IFC). In the case of the "flagship" water privatization in Buenos Aires, Suez and its partners were committed to investing up to US$1 billion in the first year. To date, it was the largest water privatization project in the world. But Suez invested only US$30 million of that money, while the rest came from the IFC and other finance institutions. The World Bank's IFC reportedly put up US$300 million plus another US$115 million to US$250 million in loans. The remaining capital came from other interested financial institutions, including US$100 million from the International Development Bank and loans from local Argentine banks. Suez has also been successful in securing IFC funding for many more of its water concessions in the South, including Sào Paulo, Brazil, and La Paz, Bolivia. Meanwhile, back in Africa, the IFC has taken the lead in an estimated US$1.2 billion project to bring in corporations to manage and invest in the water supply system of Lagos, Nigeria, and in an US$800 million water services project in Ghana. And in March 2001, the World Bank's IFC was reportedly the major overseas investor in a large water services project being developed in Thailand by RWE's new subsidiary, Thames Water International. The value of the IFC's investment was estimated to be Bt10 billion (Thai currency) (about US $225 million).

The European Bank for Reconstruction and Development (EBRD), which provides loans for both public and private sector investments in central and eastern Europe, has also provided a lot of cash that has helped the cause of the major water enterprises. The Vivendi-led consortium in the privatized Budapest Municipal Sewage Company, for example, has been given an EBRD loan of EUR$27 million to refinance its 25 percent equity stake. What this does is reduce the investment costs of Vivendi and its partners, says Public Services International, the worldwide body representing public service unions. This enhances the consortium's profit margins, but does not necessarily improve the operational performance of the water company itself. The EBRD also provided Suez with a US$90 million loan in February 2000 "to move into central and eastern European water markets" to "exploit the huge number of concession projects due to come on stream shortly." Since then, a Suez-led sewage treatment project in the Czech Republic received 70 percent of its financing from the EBRD.

Among the other regional agencies, the Asian Development Bank (ADB) has recently provided financing for numerous water privatization projects involving Vivendi and Suez. In March 2001, Vivendi secured the water and wastewater treatment concession for Tianjin, China's fourth-largest city, which will be supported by a US$130 million investment loan from the ADB (close to 40 percent of the total financing). Vivendi is also a member of a consortium in Thailand which will benefit from a US$230 million ADB loan announced in June 2001 for the Samut Prakarn Waste Management Project — an undertaking that has become the target of ongoing protests by local farmers and environmental groups. And in June 2001, the ADB announced it would be financing the lion's share (US$106 million of the US$154 million project) of the investment required for a water treatment project in Vietnam's capital, Ho Chi Minh City, to be constructed by a Suez subsidiary, Lyonnaise Vietnam Water Co.

In addition to the World Bank and its regional affiliates, the International Monetary Fund itself has recently become a major player in the financing of water privatization in the South. According to *News & Notices* prepared by the Globalization Challenge Initiative, a random review of IMF loan documents involving 40 countries revealed that the IMF imposed conditions requiring water privatization or cost recovery on 12 countries in the year 2000. Of these 12 countries, 8 are from the sub-Saharan region of Africa, generally the smallest, poorest, and most debt ridden. From the IMF's perspective, cost recovery means that everyone must be required to pay user fees to cover the full costs of the water system, which includes not only the operating and maintenance costs but also the capital expenditures. Given the close working relationships between the IMF and the World Bank, says the report, "it can be presumed that in countries where IMF loan conditions include water privatization or cost recovery requirements, there are corresponding World Bank loan conditions and water projects that are implementing the financial, managerial, and engineering details required for 'restructuring' the water sector."

Nine of the IMF loan agreements in these 12 countries were made under an IMF program called Poverty Reduction and Growth Facility. Tanzania, for example, was required to "assign the assets of Dar es Salaam Water and Sewage Authority to private management companies" as a condition for receiving debt relief from the IMF. Niger was required to privatize its four largest government enterprises (water, telecommunications, electricity, and petroleum) in its agreement with the World Bank, with the proceeds earmarked to service the country's debt payments. Meanwhile, the IMF required Rwanda to put the country's water and electricity company under private management by June 2001. In Central America, the IMF required Honduras to approve a "framework law" for the privatization of water and sewage services by December 2000, and it gave Nicaragua a "structural benchmark." This included increasing its water and sewage tariffs by 1.5 percent a month on a continuous basis for the purpose of full-cost recovery and offering concessions for private management of water and sewage systems in four regions of the country.

Yet the World Bank has had a broader role to play than even the IMF when it comes to financing water projects. Since its origins, one of the major priorities for the Bank has been the financing of hydroelectric dams in the nonindustrialized countries of the South. Between 1944 and 2000, the World Bank spent about US$58 billion dollars on 527 dams in 93 countries, according to a report published by The CornerHouse in the United Kingdom — a research agency specializing in environmental issues. As with other World Bank loans, much of the money is spent in the North rather than the South. Indeed, the World Bank's promotion of megadam projects in the South has proven to be a savior for the corporations, equipment suppliers, and technical consultants of the construction and hydro-power industries as markets have dried up in the North. "Companies meet regularly with Bank staff;" says the CornerHouse report, "invite project staff to company seminars; and involve themselves in the project cycle." Bank officials have also become particularly adept in the art of "tutoring" governments of Third World countries regarding the need to adopt dam-based development strategies.

But now, in South Africa, connections are being made between World Bank-funded mega-dam projects and privatized water services that have resulted in poor people being cut off from services and, ultimately, cholera outbreaks. According to a report published by the Alternative Information and Development Centre, an established South African NGO funded by international development agencies, the World Bank-funded Lesotho Highlands Water Project, initiated during the last years of the apartheid regime, contains two large dams (Katse and Mohale) designed to deliver more water and power to Johannesburg. For this water project — Africa's largest — the World Bank seriously overestimated the demand for Lesotho water and underestimated the cost of the dams (totalling over US$4 billion). Seven months after Nelson Mandela's election as president in 1994, World Bank officials began drafting the main sections of the Urban

Infrastructure Investment Framework for the new government. In doing so, however, the Bank refused to allow for the cross-subsidizing of water from the central government to local authorities, in order to assist poor communities. When water prices rose dramatically in the Alexandra township of Johannesburg to pay for the cost overruns on the dam projects, people in the poorer townships suddenly found themselves facing massive water cutoffs due to unpaid water bills. Without access to clean water, people in Alexandra became ill with cholera and diarrhea. On one weekend alone, four people died in that community from the cholera outbreak.

WORLD TRADE

Like the World Bank and the IMF, the World Trade Organization (WTO) has played a key role in opening up markets for transnational corporations by promoting the privatization and export of goods and services. That's why the WTO was created in 1995. In order to ensure the free flow of capital, goods, and services across national borders, the WTO has a mandate to work progressively towards eliminating all remaining tariff and non-tariff barriers. It does so by creating and enforcing an extensive body of international trade rules, including the General Agreement on Trade and Tariffs (GATT) and a battery of other trade agreements that have been negotiated with its 142 member states. In essence, therefore, the WTO promotes both deregulation and privatization. By enforcing its trade rules, the WTO makes it difficult for nation-states to either prevent imports or control exports of capital, goods, and services, including water.

Under GATT rules, water — defined as "natural or artificial waters and aerated waters" — is understood to be a tradable commodity. And Article XI of the GATT rules specifically prohibits the use of export controls for any purpose and eliminates quantitative restrictions on imports and exports. This means that if a water-rich country placed a ban or even a quota on the export of bulk water for sound environmental reasons, that decision could be challenged under the WTO as a trade-restrictive measure and a violation of international trade rules.

The same basic ruling would apply to any country attempting to restrict the importation of water as a "good" for environmental reasons. And the

WTO rules go so far as to force nations to forfeit their capacity to discriminate against imports on the basis of their consumption or production practices. Article I, "Most Favored Nation," and Article III, "National Treatment," require all WTO countries to treat "like" products as exactly the same for the purposes of trade — whether or not they were produced under ecologically sound conditions. For instance, if it were discovered that imported water had been extracted in a way that was destructive to watersheds, the receiving country might want to ban or restrict importation because of environmental concerns. However, the WTO could prevent such restrictions because any environmental safeguards or protections for water would have to be interpreted in a way that was "least trade restrictive."

Defenders of the World Trade Organization will argue that it does include an "exception" that offers some protection for the environment and natural resources like water. According to Article XX of the GATT rules, member countries can still adopt laws "necessary to protect human, animal or plant life or health . . . relating to the conservation of exhaustible natural resources if such measures are made effective in conjunction with restrictions on domestic production or consumption." However, there is something known in trade jargon as a *"chapeau"* to Article XX, which means that the Article can be applied only in a "nondiscriminatory" fashion and cannot be a disguised barrier to trade. This gives the WTO tribunals an "escape clause" that they can invoke, claiming that a particular objection, made because of environmental concerns, is actually a "disguised barrier to trade." Unfortunately, in the country dispute cases that have come before the World Trade Organization to test these "protections," tribunals have used this provision to wave aside objections based on environmental concerns. In fact, this technique has been used so many times that it appears the *chapeau* tail is wagging the Article XX dog.

In short, World Trade Organization rules are not designed to protect the environment. In all but one of the disputes that have been brought to the WTO trade panels, the rights of commerce have been upheld over the rights of the environment. Further, WTO rules trump all international environmental standards in the context of the global economy. It does not recognize, for example, the authority of Multilateral Environmental

Agreements (MEAs) in trade matters (or when it comes to trade disputes), and it threatens to undermine agreements such as the Convention on International Trade in Endangered Species of Wild Fauna and Flora (CITES). Says the U.S.-based Public Citizen, the public interest organization founded by Ralph Nader: "The emerging case law . . . indicates that the WTO keeps raising the bar against environmental laws." As a result, water conservation is at great risk under the WTO, in spite of the so-called "exception [Article XX]."

By designating water as a tradable "good" and by failing to enforce GATT Article XX, the WTO is playing right into the hands of global water exporters. The corporate players promoting water exports — delivered by pipeline, supertanker, water bag, or canal scheme, or as bottled water — can take heart in the fact that the WTO rules have been designed and are being enforced largely to protect their interests. This state of affairs is enough to wreak havoc with environmental and other responsible protections legislated by sovereign nations, but the World Trade Organization has not stopped at designating water as a tradable good. Under the WTO's General Agreement on Trade in Services (GATS), water is also considered to be a "service." Hundreds of types of water services are listed in this category — among them, fresh water services, sewer services, treatment of wastewater, Nature and landscape protection, construction of water pipes, waterways, tankers, groundwater assessment, irrigation, dams, and water transport services — to name a few.

The GATS is called a "multilateral framework agreement," which means that its broad commission was defined at its inception but that new sectors and rules are meant to be added by a process of ongoing negotiations. Initially established in 1994, the current GATS regime is already comprehensive and far reaching. Its rules apply to all modes of supplying and delivering a service, including foreign investment, cross-border provision of a service, electronic commerce, and international travel. The rules themselves are a set of legally binding constraints, designed to restrict the limits a government can put on the rights of private sector service providers to sell those services. No other international agreement to date poses such a direct threat to the legislative and regulatory power of governments. In their drafting and in their application, the GATS rules have become the power tools that the transnational corporations of the service industry need to take control of what remains of "the commons" on this planet.

The GATS rules do contain an exemption clause for services that are "delivered in the exercise of governmental authority." While this may, at first glance, appear to protect public services like water, the clause strictly refers only to *services delivered directly from governments to people*, without any commercial involvement at all. Once the private sector or the community sector is involved in the delivery of a service, or once money is exchanged, as in the case of water rate payments, then a service is no longer considered to be a government service and it is therefore not exempt from the GATS rules. As a result, most, if not all, public water systems would not be protected by this exemption.

GATS 2000

Beginning in February 2000, the WTO launched a new round of negotiations on rules governing cross-border trade in services. Known as the GATS 2000 talks, negotiations are scheduled to begin in 2002 for completion in 2005. Among the proposals under consideration is the call for an expansion of Article VI, concerning "Domestic Regulation," so that it would include a "necessity test," whereby governments would be required to prove that any measure or regulation related to maintaining a public service (like water) was "necessary." In the words of the draft article, the test would be based on "transparent and objective criteria," in accordance with "relevant international standards," and the *least trade restrictive of all possible measures*. If a government's standards on drinking water, for example, were challenged by another government on behalf of its for-profit water corporations as a barrier to trade, that defending government would have to prove that it had canvassed every conceivable way in which it might improve water quality, that its standards had been subjected to an impact assessment regarding international trade in water services, and that it had opted for the approach that was least restrictive of the rights of foreign private water providers. In other words, a government would have to go to the trouble and expense of proving that it had considered all possible private service providers (every conceivable way) that could help improve water quality and it would have to study the effects of its different possible decisions on international service providers and markets. Then it would have to opt for the approach that was most favorable to foreign private water service providers (within the context of GATS rules that favor transnationals entering domestic markets). The burden of proof, therefore, lies completely with the government being charged, not with the government and corporations bringing the complaint. With such a maze of studies to carry out and arguments to make, governments would obviously be tempted to let private companies take over their former responsibilities, just to avoid the expense and aggravation of building a complex case to defend their right to provide services through their own government agencies.

Other GATS 2000 proposals include measures to help private, foreign-based, corporate service providers in their attempts to gain access to government contracts. In this projected scenario, the WTO's existing national treatment rules on nondiscrimination would be applied to government subsidies. In other words, foreign-based, private service providers like Vivendi, Suez, and the other water mega-corporations would have the legal right to claim access to public funds for such things as government grants and loans. Another GATS 2000 proposal emphasizes the special rights of transnational corporate service providers to establish their commercial presence in another country. In contrast to goods that can always be shipped from one country to another, providing services often requires an in-country presence. Therefore, foreign service corporations, according to this proposal, would have to be allowed to invest and set up shop in another country without restriction. In short, the new GATS rules, if adopted, will give the global water lords all the legal tools they need to pry open public water systems around the world. As Canadian trade lawyer Steven Shrybman notes in his March 2001 legal opinion on the GATS: "At risk is the public ownership of water resources, public sector water services, and the authority of governments to regulate corporate activity for environmental, conservation or public health reasons."

The GATS 2000 talks have revealed, once again, that the main goal of the World Trade Organization is to protect the interests of transnational corporations at the expense of private citizens and democratic societies. However, the power of the World Trade Organization resides not only in its rules but also in its capacity to enforce those rules through the operations of its dispute settlement mechanism. Under this mechanism, member states, acting on behalf of corporations based in their countries, can challenge the laws, policies, and programs of another country as violations of WTO rules. When this happens, tribunals composed of *unelected* trade experts are given the power to adjudicate the claims and to enforce the WTO rules through legally binding mechanisms. If a country is found "guilty" but refuses to remove or change its "illegal" law, the tribunals have the authority to grant the contending country the right to invoke economic sanctions. Applied on an escalating basis, these sanctions are usually potentially damaging enough that the democratically elected government capitulates to the unelected tribunal's ruling or revises its laws in anticipation of a negative ruling.

In other words, unlike any other global institution, the World Trade Organization has both judicial and legislative powers. Its tribunals not only adjudicate disputes between countries and mete out punishments, but their rulings can have the effect of striking down domestic laws, policies, or programs that the unelected body judges to be in violation of WTO rules, or of giving rise to new ones that conform to WTO rules in advance. Although the WTO cannot directly command a member nation-state to change its laws, the threat of economic sanctions creates, if nothing else, a "chill effect" that compels governments to review and revise their legislation for fear of being targeted by a WTO tribunal.

REGIONAL BLOCS

The World Trade Organization is also flanked by regional trade regimes like the proposed Free Trade Area of the Americas (FTAA), which support international corporations in their unceasing quest for expanding global markets. Although the rules and architecture of the FTAA are currently under negotiation and are not projected to be completed until 2005, the basic framework is already in place. The future FTAA regime is being built on two existing free trade regimes in the region — the North American Free Trade Agreement (NAFTA) (covering Canada, Mexico, and the United States), and Mercosur, or the Southern Cone Common Market (covering Brazil, Argentina, Uruguay, and Paraguay). And as we shall see, the FTAA has largely been shaped in the image of NAFTA and its rules. The draft text of the FTAA, which was publicly released in July 2001, and the reports of the nine main negotiating groups that are developing the rules for "services," "investment," "market access," and other trade disciplines, make it clear that this regional regime will result in a cross-border bonanza for the water service and export giants. Like all regional trade agreements, it conforms with the World Trade Organization and its body of rules, but it also goes beyond those rules.

Like NAFTA, the FTAA will be buttressed by a dispute settlement mechanism. Unlike the requirements of the World Trade Organization, whereby corporations have to convince their home governments to take their disputes before trade tribunals for adjudication, the "investor-state" mechanism gives transnational corporations the unprecedented right to sue national governments directly, bypassing both the domestic laws and the national judicial systems of the countries being challenged. According to this mechanism, claims by corporations are adjudicated in secret by commercial arbitration panels, which can award substantial monetary damages to be paid by governments accused of violating the rules. If the FTAA is finally ratified with this mechanism in place, the transnational water lords will be in a position to directly sue any government in North or Latin America that threatens its operations, simply by claiming that it has violated certain specific trade and investment rules in the agreement.

And what are these proposed investment rules? All foreign-based water corporations would be given "national treatment" and "most favored nation" status — a state of affairs that would require any government that participates in the regime to extend to them the best type of treatment it gives to any investor, domestic or foreign. The rules are based on an extremely broad definition of investment, covering virtually all types of ownership interests. According to a proposal being advanced by the United States, the investment rules will prohibit any government from regulating inflows and outflows of capital. This allows corporations to expatriate the profits they make on a country's water services or to speculate on the sale of water rights in another country without restrictions being imposed in the public interest by the host government. The proposed investment rules are also extended far beyond the traditional definition of "expropriation" in domestic law to include "regulatory takings" — thus allowing corporations to sue any government if it passes an act or sets up a regulation that might diminish the value of their assets or profits, including future profits. (Included among such regulations would be those protecting the environment and consumer interests and those preserving public health needs.)

Although the FTAA draft is generally harmonious with the intent of the General Agreement on Trade in Services (GATS), its rules governing public services are expected to be even more sweeping. Not only will all service sectors be covered, including all water and waste management services, but the FTAA rules will apply to "all measures [defined as 'laws, rules and other regulatory acts'] affecting trade in services taken by government authorities at all levels of government." While governments will still have the right to "regulate" these services, they can do so only in ways that are compatible with the rules and "disciplines established in the context of the FTAA Agreement." So with the combined powers of the FTAA's investment and service rules, foreign-based corporations will have competitive rights to the full range of public water services in any signatory country, along with the right to sue any opposing or resisting government for financial compensation.

The unstated goal of this FTAA investment-services juggernaut is to stimulate the privatization of public services like water delivery and sanitation by reducing and destroying the capacity of governments in the hemisphere to maintain, let alone create, new public services. In outlining rules for "domestic regulation," for example, the draft FTAA text proposes its own version of the "necessity test" by calling on governments to "limit the regulation in scope to what is necessary" and to "avoid unnecessary regulations." These wordings have disturbing implications. Government regulations requiring certain water quality standards, lower water rates for the poor, or improvements in pipe infrastructure could be declared as "unnecessary" by an investor-state tribunal under the FTAA. Instead of allowing governments to defend the interests of their citizens by regulating the operations of corporations through laws, policies, and programs, the FTAA calls on governments to "stimulate the use of market mechanisms for regulatory objectives." In other words, give corporations more tax breaks as an incentive to act in the public interest.

The main problem with such a scenario is the fact that there is no assurance that transnational corporations will respond to favorable market conditions by being socially responsible toward the people of the countries in which it is operating. The general public is left to the mercy of a transnational corporation that may not feel any moral compulsion to act in the interests of the people it is billing for water services. Given a favorable business climate, created by tax breaks or other "market mechanisms for regulatory objectives," a corporation may just as likely decide to keep the extra profits for itself as to repair a defect in the way it is delivering services. Traditionally, in cases where moral compulsion is considered to be an insufficient incentive, socially responsible behavior is encouraged through legislation and the threat of punitive measures if such legislation is not adhered to. The proposed FTAA regime will dismantle this longstanding democratic tradition. As a result, both citizens and communities are left in a vulnerable situation, as corporatins are governed by the laws of the market alone.

The FTAA measures will also be reinforced by the agreement's market access rules, which will lay down a schedule for governments to eliminate all their tariff barriers (e.g., border taxes) and especially non-tariff barriers to free trade. Non-tariff barriers include all the laws, policies, and practices of governments that inhibit cross-border trade. This can include anything from a government's delivery of water services to its efforts to protect public health and safety. Indeed, when it comes to public services, government regulations are considered to be "non-tariff barriers." For instance, if a water system being delivered by a government is deemed to be a "monopoly," the "national treatment" rule may be used to declare the public water monopoly "discriminatory" against foreign-based water corporations seeking market opportunities for their services.

If the FTAA adopts provisions on natural resources similar to those that already exist under NAFTA, it will greatly enhance the powers and tools available to transnational water exporters as well. Simply put, NAFTA prohibits any government from putting a ban on the export of natural resources, including water. Article 309 of NAFTA specifically states that "no party may adopt or maintain any prohibition or restriction on the exportation or sell for export of any good destined for the territory of another party." This also implies that no government is permitted to collect export taxes on water shipped beyond its borders. In addition, NAFTA includes a "proportionality clause" (Article 315) which specifies that the government of a member country cannot reduce or restrict the export of a resource to another member country once the export flow has been established. In other words, if water exports between Canada and the United States or Mexico were to begin, the tap could not be turned off and the flow could not even be reduced from previous export levels. Instead, exports of water would be guaranteed at the level established over the preceding 36 months.

More than likely, a variation of these NAFTA export rules will be incorporated into the FTAA, especially given President George W. Bush's dream of setting up a continental energy and water corridor. And once these export rules are in place, they cannot be revoked by a member country, even if new evidence is found that extractions of bulk water were destructive to the ecosystem. If Brazil, for example, were to ban bulk exports of water for environmental reasons under an FTAA that included these export rules on natural resources, it could be directly sued by a water export corporation through the investor-state mechanism. Similarly, if the state of Alaska were to reverse its policy and ban water exports or change its law so that only U.S. companies could export Alaskan water in order to keep jobs at home, the U.S. could be slapped with an investor-state suit by a Canadian-based company like Global H2O, which has a contract to ship 18.2 billion US gallons (about 69 billion liters) of glacier water from the Alaskan town of Sitka to China.

So far, there has been one major suit filed against a government ban on water exports using NAFTA's investor-state provisions. In the fall of 1998, the Sun Belt Water Corporation of Santa Barbara, California, sued the Canadian government because the company lost a contract to export water to California when the Canadian province of British Columbia banned the export of bulk water in 1991. Sun Belt claims that the B.C. ban contravenes the investment and export rules of NAFTA and is seeking US$10 billion in damages. "Because of NAFTA," declared Sun Belt's CEO Jack Lindsay, "we are now stakeholders in the national water policy in Canada." These water export challenges are bound to multiply in the near future, especially if predictions of water shortages in places like the United States and Mexico intensify and if more serious environmental concerns arise around the issue of bulk water extractions. And if similar export rules are incorporated into the FTAA, the number of potential challenges will only increase.

The strongest impact of these provisions, however, may not come through lawsuits filed by corporations. The very existence of these rules, fortified by the threat of lawsuits permitted under the investor-state mechanism, is sufficient to cast a "regulatory chill" over government policies and legislation. Faced with the prospect of being slapped with a multimillion- or multibillion-dollar suit, governments tend not to go ahead with putting a new law or regulation on the books if there is a chance that it might be challenged as a violation of these trade regimes. Increasingly, governments are reviewing all proposed laws and regulations in light of a "trade screen test" before they can be enacted by their elected legislatures. In short, trade regimes like the FTAA serve to transfer political power from governments to corporations, allowing them to move to the most lucrative markets in the Americas whenever they want. It is difficult for governments to oppose them, since they are armed with a legal tool box that can compel jurisdictions to commodify and privatize public services, including public water systems.

At the same time, the International Monetary Fund and World Bank "Structural Adjustment Programs" (SAPs) are undermining the ability of sovereign governments to act according to democratic principles. After all, the SAPs were the financial tools used to compel debtor countries in the South to integrate themselves into the global economy by reducing their public sectors; slashing government spending for health, education, and social services; privatizing state enterprises; and redirecting their domestic economies toward export-oriented production. Over the past two decades, the SAPs have already set the conditions for the corporate takeover of water systems in the Americas. What the FTAA will do, reinforced by the WTO, is to provide the global water corporations with the legal tools to further their privatization and export agendas.

INVESTMENT TREATIES

The global commons, including water, are also under threat of being commodified and concentrated in the hands of a very few because of mechanisms like investment treaties. In the early 1960s, Germany and France, among others, began concluding Bilateral Investment Treaties (BITs) with a number of countries. For the most part, these BITs were designed to establish the investor rights of corporations from each country to operate unconditionally in the other country and to access each other's markets and resources. Since 1994, many of the BITs negotiated between countries began to incorporate some of the key rules and disciplines of NAFTA, including the rights of investors, the broad definition of investment, restrictions on government-initiated performance requirements such as export quotas, direct and indirect forms of expropriation, and the investor-state mechanism itself.

According to the United Nations Conference on Trade and Development (UNCTAD), 1,310 BITs had been signed worldwide by January 1997, of which the largest number were concluded by western European countries. By 2001, however, the number of BITs had risen to 1,720, showing a steady growth rate year by year. But in spite of their widespread nature, these BITs are among the "best-kept secrets" in the international community. Very few politicians, let alone citizens, are aware they exist, and fewer have any knowledge of their contents and the powers they grant to transnational corporations operating outside their own countries. However, these BITs can provide global water corporations with the additional economic and political clout they need to pry open markets and resources, if the governments of countries they are already operating in have BITs with other nations in which they want to carry out business. And that power is all the greater if those BITs contain the investor-state mechanism that allows them to directly sue the government of the host country.

This is what Bechtel, for example, has done in retaliation against the Bolivian government for canceling its water contract in Cochabamba after mass resistance in the streets. Under a 1992 BIT signed between Bolivia and the Netherlands, Bechtel is using one of its Dutch holding companies to sue the Bolivian government for US$40 million in compensation for "expropriation" rights, due to "losses" incurred by the cancellation of the water contract in Cochabamba. Bechtel, of course, is a U.S.-based corporation, but in 1999 it moved its Dutch holding company for Aquas del Tunari from the Cayman Islands to the Netherlands, thereby gaining the right to sue Latin America's poorest country at the World Bank's International Centre for Settlement of Investment Disputes. Since Bechtel's actions became known in November 2000, the Bolivian government has publicly stated that it will fight this challenge. But there are others in the government who feel it would be best to pay Bechtel its compensation demands in order to prove that Bolivia is ready for economic globalization and can be a "good" player in the new world order dictated by the WTO. As a result, there are real concerns that Bolivian government officials are engaged in closed-door negotiations to settle the dispute out of court.

In another case, Vivendi has made use of a BIT between its home country, France, and Argentina to bring a suit against Argentina and the government of the province of Tucumán. In 1995, Vivendi's water company, Générale des Eaux, and its Argentinian affiliate Compañia de Aguas del Aconquija signed a concession contract with Tucumán province to manage its water and sewage facilities. When the provincial government's health authorities imposed fines on the companies for failing to install proper water-testing equipment, and the provincial ombudsmen denied the companies the right to cut off water services to nonpaying customers, and the government refused to allow rate increases to be charged, Vivendi filed a complaint with the French government through its water subsidiary and affiliate. After attempts by the governments of France and Argentina to resolve the dispute, Vivendi made use of the BIT, which contained features similar to NAFTA, to bring a US$300 million claim against Argentina. The dispute has sparked a considerable amount of public debate and controversy in both countries. Upon hearing the case, a tribunal instructed Vivendi to pursue the matter through the courts in the province of Tucumán first, before seeking recourse to international arbitration.

For transnational corporations like Vivendi and Bechtel, Bilateral Investment Treaties have become increasingly important tools since the defeat of the Multilateral Investment Treaty (MAI) in 1998. Initially drafted by the International Chamber of Commerce in 1996, the MAI was negotiated under the auspices of the Organisation for Economic Co-operation and Development (OECD), known popularly as the "rich nations club." The MAI would have given corporations sovereign powers over nation-states, including the legal tools to dismantle public enterprises (or what it called "state monopolies") through commercialization; prohibit governments from putting any bans or quotas on exports of natural resources; roll back laws, policies, and programs that did not conform to the MAI's investment rules; and enforce the MAI's rules and disciplines through an investor-state mechanism that would have allowed foreign-based corporations to sue governments directly. The MAI was a major issue in the early stages of the movement against corporate globalization. Mounting public opposition, coupled with splits between several of the European Union countries and lukewarm support from the United States ultimately led to the collapse of the MAI negotiations at the OECD in October 1998. Had the MAI been ratified as an international treaty, however, it would have given corporations, including the water giants, both the constitutional and the legal powers to exploit what remains of the global commons. Originally called the "Constitution for the Global Economy" by the director general of the World Trade Organization, the MAI would have had the power to override the constitutions of many nation-states. If it had been adopted, constitutional amendments would have been required to make a country's constitution compatible with MAI provisions.

Despite the MAI setback, the nexus between corporations and governments around the world has since proceeded apace. In July 2000, the United Nations announced a "Global Compact" with a number of well-known transnational corporations that had agreed to adopt certain social responsibility guidelines on a voluntary basis. The list of corporations included notables like Shell, Nike, and the water giant Suez. After all, the public debate surrounding the Multilateral Investment Treaty and its eventual defeat, combined with growing opposition to institutions of corporate globalization like the WTO, the IMF, and the World Bank, made it clear to many transnational corporations that they were losing moral and political legitimacy as well as public support. What's more, the Global Compact facelift could be done at relatively little or no cost. In contrast to trade regimes like NAFTA and the proposed FTAA, in which investor rights had become binding and enforceable, the social responsibility guidelines in the Global Compact were completely voluntary and therefore ineffective as a counterweight.

Then, at the November 2001 Fourth Ministerial meeting of the WTO in Qatar, a section on "Trade and Environment" that endangers freshwater was added at the last minute by Europe (home of the big water companies). It calls for "the reduction, or, as appropriate, elimination of tariff and non-tariff barriers to environmental goods and services." Water is a "service" already in the GATS and a "good" in the GATT. Soon, if the WTO gets its way, it will be an "invesment" as well. Under this dangerous new WTO provision, a domestic rule that protects water as a public service and a human right could be considered a "non-tariff barrier" to trade and eliminated, as could any rule that attempts to limit privatization. Further, says the new text, domestic "environmental services" (water regulation) must be "compatible with" other WTO rules, such as those that prevent nations from using "non-tariff barriers" such as environmental laws, in a way that interferes with trade liberalization. Domestic environmental standards to protect water could be challenged by subsuming the trade in environmental services under these already dangerous WTO agreements.

In spite of these attempts by corporations to increase their perceived legitimacy, citizens of many countries still consider that transnationals are more interested in maximizing profits than in serving the common good. As groundwater is siphoned out for exports, local wells go dry. When public services have been privatized in Third World countries, rates have skyrocketed and the poor have been cut off from steady water and sewage services. And corporations around the world are using clauses in Bilateral Investment Treaties and the WTO, NAFTA, and GATS, as well as the IMF and the World Bank's Structural Adjustment Programs to retaliate against governments that try to impose penalties for not acting according to the standards of that government. As citizens around the world observe practices like these, which diminish the power of democratic governments, they have begun mobilizing to resist massive privatization and to gain back control of the commons.

PART III

THE WAY FORWARD

FIGHTBACK

*How people around the world
are actively resisting the
theft of their water rights*

For the third time in 1999, the flood waters rose in the Narmada Valley of India following another phase in the construction of the Sardar Sarovar Dam. Once again, the villagers and activists stuck to their long-standing pledge of refusing to move to resettlement sites, even if that meant they would drown in the face of the rising waters of the reservoir.

The villagers and their supporters were part of a people's movement known as Narmada Bachao Andolan (NBA), which united groups affected by the construction of three mega-dams — the Sardar Sarovar, the Narmada Sagar, and the Maheshwar — in the Narmada Valley. The NBA was initially inspired by a woman named Medha Patkar, who came to the Narmada Valley to study the areas that were to be submerged by the Sardar Sarovar Dam and its river-killing reservoir system. After witnessing the extensive ecological and social harm that could be caused by the project, Patkar decided to spend her time traveling throughout the proposed submergence zone, meeting with the people who were to be displaced and urging them to organize.

Analyzing the official documents, Patkar and her associates alleged that key environmental impact studies had not been conducted; that there was no knowledge of the numbers of people who would be displaced; that estimates about the amount of land that would be irrigated by the project were greatly exaggerated; and that the funds for building the water supply system, which had been one of the main features of the project, were not even included in the financial assessments. From 1990 onwards, the NBA pledged to actively resist any dam construction through nonviolent direct action until there had been an independent, participatory review of the project and its impacts.

One of the crucial battles waged by the NBA over the past decade occurred in 1991 when thousands of villagers and activists joined in a "Long March" and a 21-day fast, demanding that the World Bank, which was funding the Sardar Sarovar Dam, commission an independent review of the whole project. With the glare of the international media spotlight on them, the World Bank agreed to the demand. The independent review, resulting in the *Morse Report*, declared the project to be environmentally flawed and was critical of the roles of the World Bank and the government of India.

When the government refused to meet certain minimal conditions, the Bank took the unprecedented step of canceling its financial role in the project. Determined to complete the dam construction, however, the government was able to scrape together financing from other sources. Then, between 1993 and 1995, during the monsoon season, police repeatedly arrested villagers in the lowest-lying houses, who refused to move as the flooding escalated, and dragged them to higher ground. Finally, in early 1995, the Supreme Court of India halted all construction on the dam, in response to a case filed by the NBA. The injunction continued until 1999, when the government persuaded the court to allow the Sardar Sarovar dam to be raised several meters, thereby causing more floods and a new round of resistance.

⌣

During the past decade, the anti-dam movement led by the NBA in the Narmada Valley has become a symbol of the people's struggle for water rights around the world. In addition to opposing the Sardar Sarovar, villagers have organized against the construction of mega-dams through-out the valley. In January 2000, for example, local villagers took over the Maheshwar Dam for the eighth time in three years. The resistance organized by villagers in the Narmada Valley, however, is not simply moti-vated by the displacement and resettlement caused by the construction of dams. They are also protesting because these mega-dams destroy their traditional water-harvesting systems by creating river-killing reservoirs. The Narmada villagers know full well that they have a fundamental right to water as the essence of life. They also know that water can be provided from the valley for irrigation and drinking needs without constructing huge dams and destroying natural river systems.

Although anti-dam protests have been at the forefront of the fight for water rights, people all over the world have become increasingly involved in a wide range of community struggles to stop the theft of their water. These battles are being fought in local communities over a variety of issues, from the privatization, export, and quality of water to the preservation of lakes, rivers, and watersheds. At the center of these community-based struggles is a growing resistance to the commodification of water resources and the corporate takeover of water systems.

PUBLIC CONTROL

One of the more significant kinds of water fightback stories taking place in recent years has to do with communities that have struggled to regain public control of their municipal water services after they have been privatized. In particular, two stories that we have touched on previously stand out. The first story comes from the recent battle against water priva-tization in Cochabamba in Bolivia; the second is based on a determined effort over several years to turn the tables on the private control of water services in the city of Grenoble, France.

As we saw at the beginning of Chapter 7, La Coordiadora de Defense de Agua y la Vida, led by Oscar Olivera, was successful in mobilizing widespread resistance in Cochabamba, Bolivia, after the government, responding to conditions stipulated by the World Bank, agreed to have a subsidiary of the Bechtel Corporation take over the city's public water system. Organized as a broad-based movement of workers, peasants, farmers, and other concerned citizens, La Coordiadora's main goals were to "de-privatize" the local water system and to defend the community's rights to "water and life." When Bechtel's privatization of Cochabamba's water system sent water bills skyrocketing and the government was forbidden to use World Bank loan monies to subsidize water services for the poor, Bolivians by the thousands marched to Cochabamba. They provoked a general strike and transportation blockades that brought the city to a standstill. By the time the government declared martial law in early April 2000, the police were reacting to the mass protests with violence, activists were being rounded up and arrested at night, and radio and TV programs were being shut down in mid-program.

This mass resistance, led by La Coordiadora, was successful both in compelling the Bechtel subsidiary to leave the country and in convincing the Bolivian government to back down on privatization. On April 10, 2000, the directors of Bechtel's subsidiary, Aguas del Tunari, packed up and left Bolivia. Under intense popular pressure, the Bolivian government rescinded its detested water privatization legislation. But now there was no one left to run the local water company, Servicio Municipal del Agua Potable y Alcantarillado (SEMAPO), so the Bolivian government handed Cochabamba's water services over to the workers at SEMAPO and the community itself.

Accepting the challenge, the community set out to elect a new board of directors for the water company and develop a new mandate, based on a set of alternative principles, which are still in force. SEMAPO's new principles state that the company must operate in a manner that is efficient, free from corruption, and fair to its workers. Guided by a commitment to social justice, the company must be prepared to provide first for those without water and act as a catalyst to further engage and organize the grassroots. Inspired by this mandate, the new company immediately set up a huge water tank to serve the poorest areas of Cochabamba, developed links with four hundred small communities previously abandoned by the old company, and began working directly with local neighborhoods to solve water service problems. By the summer of 2000, La Coordiadora itself had organized the first of a series of public hearings to begin the process of building a broad-based community consensus as to what the water company should become in the future.

Meanwhile, on the other side of the Atlantic Ocean, the citizens of Grenoble in France — the national home of the world's leading water giants — were celebrating the return of their water service and sewage system to public control after more than a decade of local community struggles. In 1989, the mayor of Grenoble had initiated proceedings to privatize the city's water services by striking a deal with Lyonnaise des Eaux, a subsidiary of that world-leading water company, Suez. Despite strong public opposition, the contract to privatize went ahead. As we discussed in Chapter 5, however, the deal was riddled with corruption. The privatization scheme was concluded in exchange for monetary contributions to the mayor's electoral campaign. When Lyonnaise des Eaux introduced massive hikes in water prices, public resistance grew and a citizens' movement was born. Then, in 1995, the mayor and an executive of Lyonnaise were prosecuted and in 1996, they were convicted of bribery.

The citizens' movement that emerged was anchored in two organizations: the Association for Democracy, Ecology, and Solidarity (ADES), and Eau Secours ("Save the Water"). Both organizations went to work, doing background research on the deal that was struck with the Suez subsidiary. Eventually, both ADES and Eau Secours put together a legal strategy to challenge the water privatization deal in court. Through these initiatives, the citizens of Grenoble won a series of court rulings that overturned the price hikes. The courts also went on to nullify the original 1989 privatization decision and subsequent contracting out of the city's water and sewage system. Following this court action, the Grenoble city council opted for the creation of a "*société mixte*" and proceeded to subcontract its water services to another subsidiary of Lyonnaise des Eaux. But this contract was also declared null and void by a court ruling instigated by yet another legal challenge brought by the citizens' movement.

The stage was now set for the de-privatization of Grenoble's water system. Since 1995, citizen activists had been waging electoral campaigns based on a platform of returning the city's water system to public hands. After winning several seats on council, the first day of spring in the new millennium gave the citizens of Grenoble something to celebrate. After a decade of privatization, Lyonnaise des Eaux was handed its exit papers. In March 2000, the Grenoble city council decided to return the water and sewage system to public control once and for all.

FIGHTING PRIVATIZATION

The stories of Cochabamba and Grenoble illustrate what people can do when they organize to take back public control of their water systems. And the fight against the privatization, or corporate takeover, of community water systems has been accelerating recently in countries all over the world. This process has been assisted by organizations like Public Services International (PSI), the worldwide alliance of public service unions, and its affiliates.

In South Africa, the only country in the world where people's right to water is actually written into the national constitution, the townships surrounding cities like Johannesburg and Durban have recently become hotbeds of resistance to water privatization. Labor unions like the South Africa Municipal Workers' Union have been openly challenging the privatization plans of Suez, Bi-Water, and other water lords and actively promoting a model of "public-public partnerships," as an alternative to privatization. At the same time, people in the poor townships of Johannesburg are being faced with water cutoffs as a result of not being able to pay for rising rates charged so the water companies can attain full cost recovery. These citizens have been organizing resistance on a neighborhood-by-neighborhood basis. When cutoffs occur, local teams go from household to household, hooking up the water supply again and pulling out the water meters. Similarly, when water services were cut off in the Township of Empangeni on the outskirts of Durban, citizens rebelled by pulling out the water meters. In May 2001, people from Empangeni Township also squatted on government land to hold their own conference on the water crisis, after the mayor denied them the use of any public facility for their meeting.

In Ghana, where the IMF and World Bank have insisted on water privatization as a condition for renewal of loans, a broad range of civil society organizations have formed a National Coalition against the Privatization of Water. In response to reports that 44 percent of the Ghanaian population have no access to water services, the coalition issued the "Accra Declaration on the Right to Water" on June 5, 2001. At the heart of the Accra Declaration was a rejection of the commodification of water and the model of privatization, driven by foreign-based transnational corporations, as "the appropriate solution to the problems bedeviling our water sector." The declaration called on the government of Ghana "to reverse the decision to put the privatization process on a fast track"; to investigate alternative models of water-service delivery that make greater use of local communities, governments, and businesses; and to conduct "a country-wide debate on options for reforming the water sector." The coalition itself is committed to a program of action that includes a broad-based campaign to ensure all Ghanaians have universal access to water by 2010; constitutional guarantees for people's right to water; public ownership, control, and management of water services; and promotion of alternative solutions to problems of public management efficiency.

On the other side of the Atlantic, in Uruguay, a coalition of labor unions and social organizations calling itself the Movement for a Popular Initiative (MPI) has been organized for the purpose of developing laws to prevent further privatization of water services. In a 1992 referendum, 70 percent of the Uruguayan people said "no" to the privatization of public services. By January 2000, however, the Spanish water corporation Agua de Barcelona had been granted a 30-year concession to manage the water and sewage systems in Montevideo and other communities. Whereas citizens have had a choice between using a public or a private sector sewage system in the past, now the private system is compulsory. A subsidiary of Suez has also been granted a permit to develop a database on Uruguay's abundant supplies of groundwater. Given that Uruguay's legislative process includes a provision allowing citizens to take their own initiative in making laws, the MPI developed a draft law in 2001 that would require the government to prevent any further privatization and to abrogate the current budgetary provisions for privatization of the water system. According to this provision, if the government does not accept this citizens' initiative, then it is to be put to the people for a referendum.

In the United States, community groups, public service workers, and municipal councilors have often worked together to resist water privatization in cities and towns throughout America. One of the common targets has been the American Water Works Company. For instance, the town of Pekin, Illinois, after several years of frustration, took steps to buy back its water system from American Water in 1999. After spending two years in court fighting American Water, Huber Heights, a suburb of Dayton, Ohio, won its case and went on to receive a 75 percent vote from its residents to buy back its municipal water system in 1995. Similarly, the residents of Birmingham, Alabama, turned down a US$390 million offer from American Water in 1998 to buy their water system, while the people of Nashville, Tennessee, took steps to ensure that their water system would remain in public hands. Even in Orange County, California, which had to declare bankruptcy in 1994 after losing US$1.7 billion on the stock market, the Santa Margarita Water District said "no thanks" to a US$300 million offer from American Water to buy its water system, preferring to keep it in public hands.

Similar struggles have been waged across Canada to protect public water and sewage systems from corporate takeover and control. In Vancouver, a thousand people jammed a public hearing of the Greater Vancouver Regional District (GVRD) in June 2001 to demand that proposals for the privatization of the Seymour filtration plant, which included subsidiaries of Vivendi and Bechtel among its bidders, be withdrawn. The public opposition, mobilized by groups like the Canadian Union of Public Employees (CUPE), the Council of Canadians, and the Society Promoting Environmental Conservation, emphasized that once the water treatment system was privatized, trade regimes like NAFTA and the WTO would prevent the city government from reclaiming public control in the future. In response, the GVRD promptly canceled its water privatization plans, which reportedly had been in the works for three years.

In Kamloops, British Columbia, citizens' groups and public service workers were successful in preventing the city government from proceeding with plans to enter into a public-private partnership for building a new water treatment facility. And on the other side of the country, municipal leaders in the coastal communities of Newfoundland pledged to improve their sewage treatment facilities, which were affecting the St. John's harbor.

In this case, CUPE met with the mayors of the area to urge them to retain public ownership and control. Despite pressure from business leaders in St. John's, the mayors have so far refused to entertain bids from the big water corporations.

WATER EXPORTS

Although widespread resistance has been organized against the damming of river systems and protests against the privatization of water systems are on the rise, the fightback against water exports is only in the early stages of development. In part, this may have to do with the fact that many of the schemes for bulk water exports through pipelines and canals and in supertankers and water bags are still in the experimental and planning stages. Bottled water, however, is another story, and there are signs that opposition is beginning to emerge on this front.

Nestlé's leading brand of bottled water, Perrier, for instance, has become the focal point of popular resistance in the state of Wisconsin in the United States. With a pumping permit granted by the state's department of natural resources, Wisconsin groundwater has become Perrier's main source of spring water for its Ice Mountain brand. A group called the Concerned Citizens of Newport has been leading a public fight to prevent the mega-pumping necessitated by this business and to stop the degradation of associated wetlands. "Taking spring water out of any ecosystem is like taking blood out of people," says activist John Steinhaus. In two town referendums, the people firmly rejected Perrier's pumping of local spring water by ratios of 4 to 1 and 3 to 1. After repeated testimonies before state senate hearings and public information meetings, Concerned Citizens filed a lawsuit against the Wisconsin Department of Natural Resources in October 2000.

Similarly, in July 2001, the Michigan Citizens for Water Conservation (MCWC) launched a protest against a petition by the Perrier Group of America for a state license to proceed with a project involving the privatization, diversion, and export of water from a portion of the Great Lakes. By installing "a high-capacity non-community private water well system" the Perrier Group planned to extract spring water at a rate of 17 million US gallons (64 million liters) a month or 204 million gallons (772 million liters) a year — enough to fill a 12-acre (5-hectare) lake to a depth of 51 feet (about 15 meters) every year. The MCWC contends that this project "has the potential to do serious long term damage to the area's water and natural resources, and to seriously compromise Michigan's ability to protect the waters of the Great Lakes." According to the MCWC, the Perrier project will also diminish the surface waters of Osprey Lake and the stream below it, and the pumping operations will adversely affect over 40 acres (16 hectares) of wetlands. Citing the Michigan *Environment Protection Act*, the *Inland Lakes and Streams Act*, and the *Wetlands Protection Act*, the MCWC insists that the Perrier project must not be licensed or approved by Michigan state authorities.

Many more local citizens' campaigns are likely to be launched to combat the extraction of spring water for export — especially as PepsiCo and Coca-Cola battle more intensely over the bottled water market. But at the moment, one of the main areas of contention is sources of water located in Aboriginal communities. According to a consulting firm report prepared for the Canadian government and obtained under the *Access to Info   'ion Act*, investment brokers have been making multimillion-dollar offers to Native organizations to extract and export water from lakes and rivers on treaty lands. Based on the strong links between water and life which are deeply embedded in the cultural and spiritual traditions of most Aboriginal communities, the Assembly of First Nations in Canada has taken a clear stand against water exports and joined with other civil society organizations in calling on the Canadian government to ban bulk water exports. Yet as the consultants' report points out, there are "concerns that poor communities were being offered multimillion dollar deals that would be hard to resist." Without an export ban, many communities may very well succumb to the temptation.

In Canada, the whole issue of bulk water exports has already become a focal point for public fightback campaigns. When the McCurdy Group proposed its scheme to extract and ship bulk water from Gisborne Lake in Newfoundland and then ship it by supertanker to the Middle East, the Council of Canadians mounted a mass protest. Public rallies were organized in October 1999, and meetings were held with the Newfoundland government, which, in turn, promptly announced it would not grant the export license. Although the resistance has since tapered off, it is bound to ignite again, given recent statements by the province's new premier indicating that Newfoundland may be poised to grant the McCurdy Group the license to go ahead after all. Meanwhile, the Council has also been leading a national campaign calling on the Canadian government to pass legislation banning bulk water exports. While recognizing strong public support for the Council's position on the issue, the Chrétien government has so far refused to take this legislative initiative, opting for an accord between the provinces against water exports instead. By failing to take leadership itself in banning bulk water exports, warns the Council, the government is simply allowing trade regimes like the WTO and NAFTA to dictate Canada's policies.

Elsewhere, the Council has played an active role in legally challenging proposed schemes for bulk water exports from sources like Tay River, near Perth, in southeastern Ontario, as well as building public support for the export ban issued by the former New Democratic Party government in British Columbia. But Canada is certainly not the only potential hotbed of resistance to bulk water exports. Environmental movements in Europe, for instance, have been mounting campaigns against bulk exports of water from the Austrian Alps and from Norway. And as more evidence of the ecosystem damage caused by bulk water extractions becomes known, the public opposition on this front is bound to intensify.

WATER QUALITY

Unlike the battle against water exports, the fight against water contamination has been going on for some time. From the extensive use of chemicals in agribusiness to waste disposal from oil, gas, and mining projects, megabusiness activities are polluting natural water systems on a massive scale. And this has given rise to a loose network of fightback struggles for water quality in communities throughout the world. These actions have taken form in diverse ways.

In Colombia, for example, a coalition of environmental, peasant, worker, and human rights groups has been confronting the Occidental Petroleum Corporation on issues of water contamination since the mid-1990s. Occidental's big oil production and pipeline facility, Caño Limon, was built on a flood plain in 1986, with devastating consequences for the natural water systems in the region. When heavy rains fall, says the coalition in its report, they wash over the open pits, carrying both toxic and carcinogenic chemical waste into local waterways like the Arauca River. The alarm was first sounded in 1988, when the Institute of Natural Resources in Colombia (known as INDERENA) disclosed that the type of water-treatment facility used at Caño Limon would have been environmentally unacceptable in the United States because it fails to protect surface waters from contamination. It is also forced to handle far more wastewater each day than its maximum capacity, and there is no mechanism for dealing with toxic muds. In a 1992 environmental study, the institute found dangerously high concentrations of heavy metals and toxic hydrocarbons in the local water system, three hundred times higher levels than drinking-water standards in the North. In 1998, a lawsuit was brought against Occidental Petroleum for environmental damages, including contamination of local water supplies.

Closer to home, a group of farmers, ranchers, and concerned citizens in Alberta was formed in 1999 to challenge the oil industry's heavy use of fresh water aquifers to "pressure-up oil fields" so that a larger percentage of the oil can be extracted from the field. With this method, water is injected into the earth to the depth of an oil well, thereby creating an "oil flood." When this happens, water is not only removed from the water cycle but it is also completely contaminated in the process. In 2000, it is estimated that water-diversion permits for oil floods in Alberta totaled 45.4 billion Imperial gallons (about 206 billion liters), with 17 billion gallons (about 77 billion liters) coming from underground aquifers alone. In response to a Petro-Canada application for a permit to draw water at depths of 0 to 500 meters (about 1,600 feet) for a large oil field, the group initially mobilized protest letters, petitions, and actions. Although failing to stop the Petro-Canada permit, the group decided to become a clear public voice of opposition to the "misuse and abuse of water by oil producers." Through a variety of actions, the group has begun to publicly challenge Alberta's Department of Environment for promoting the interests of the oil industry, rather than conserving and protecting water quality, and the group has also exposed the waste and contamination of fresh water supplies by specific petroleum corporations.

At the same time, the heavy use of chemical fertilizers and pesticides in industrial agriculture has become a major cause of ongoing water contamination and a focal point for protest. With rainfall, toxic and carcinogenic substances in farm chemicals are washed through the soil and into surface and underground water systems, polluting and poisoning the water supply. This has been one of the key issues in campaigns organized by the Pesticide Action Network (PAN). As an international coalition, PAN is a network comprising over 400 organizations working in 60 countries. With hubs in Africa, Asia-Pacific, Europe, and North and Latin America, PAN has mounted numerous campaigns focusing on the dangers of using chemical pesticides in agriculture, including the contamination of groundwater systems. Through its campaigns, PAN provides opportunities for farmers, farm workers, and communities to learn about different chemicals, their uses and abuses, and what damage they can do to the ecosystem and biodiversity, as well as to humans (as a cause of cancer, birth defects, or neurological damage). As an alternative method of agriculture, PAN consistently promotes organic farming, which would have much more positive effects on water quality — among other benefits.

In recent years, the proliferation of industrial hog farms has also become a major issue for water-quality campaigns. As president of the Water Keeper Alliance in the United States, Robert F. Kennedy, Jr., launched a major legal assault on America's large pig factories in December 2000. In order to confine large numbers of hogs in small crates in these factories, one of the methods used is to deny them bedding so that their manure can be liquefied for easy handling. This liquefied manure, in turn, is allowed to run into streams and seep into groundwater, emitting toxic gases. Calling this a monumental environmental crime, Kennedy and the Water Keeper Alliance have charged the U.S. government with failing to enforce environmental protection laws in relation to the industrial hog farms. Similar campaigns are also being organized in Canada by farmers and citizens in rural communities who fear that their groundwater is seriously threatened by toxic wastes generated by these modern pig factories. What's more, these factories are cruel "animal concentration camps," where sows are not even able to walk or turn around.

As we saw in Chapters 1 and 4, the high-tech computer industry has also become a key target in campaigns against water pollution in places like Silicon Valley and Phoenix, Arizona. During the past decade or so, groups including the Silicon Valley Toxics Coalition, the Southwest Network for Economic and Environmental Justice, and the Campaign for Responsible Technology have been organizing community resistance to the constant polluting and poisoning of local water systems by the high-tech computer industry. In part, their public education campaigns have served to unmask the hypocrisy of this sector, which claims to use "clean" production methods, but which, in fact, has a staggering pollution legacy. These campaigns have also exposed the extent to which government agencies favor the interests of high-tech companies by failing to rigorously enforce environmental protection and cleanup legislation as it applies to this industry.

Citizen activists and public service workers have also been emphasizing water quality issues in their anti-privatization campaigns. In its fightback against the corporate takeover of public filtration and sanitation systems, for example, the Canadian Union of Public Employees has found that people are showing much greater concern about issues of water contamination and public accountability. In particular, small food vendors and restaurant owners have become more sensitive about the need to protect water quality. If our water filtration systems are put in private hands, they ask, where is the public accountability? "Handing over the management of our water treatment system to a private firm just doesn't make sense," says a deli store owner in Kamloops, British Columbia. "Why would we jeopardize our health by placing responsibility for water-quality standards in the hands of a corporation that answers to its shareholders rather than city residents."

RESTORING WATER SYSTEMS

Fighting back can take on many different forms — including protecting individual water systems or whole watersheds (the areas drained by a particular group of lakes, bays, rivers, and streams). As with the other battles, people are sometimes taking control themselves — in this case, working together to restore, save, or revitalize rivers or watersheds and thus protecting them for future generations.

In 1990, for example, a group of American environmentalists known as Ecotrust worked to save a pristine watershed in northern British Columbia. In the United States, they had been actively involved in fighting for the preservation of rainforests in other countries, and in this battle, they joined hands with the Haisla First Nation, whose ancestors had lived since time immemorial in the Kitlope Valley. The valley itself is blessed with lakes, rivers, streams, and a rich flood plain — and all six species of salmon swim the waters of the Kitlope River. After preparing an ecological inventory of the Kitlope watershed, in 1992 Ecotrust and the Haisla worked with a group of Native and Non-Native residents in the area to develop a wilderness-planning framework. Outlining alternatives to logging and resource extraction, the framework proposed a sustainable economy based on guided tourism, ecosystem research, wildlife viewing, and rediscovery camps. To implement this plan, the Haisla and Ecotrust set up the Nanakila Institute (using the Haisla word meaning "to guard; to watch over") in 1993 to train people from the Aboriginal community to work in these areas. Eventually, this collective work became the basis for negotiating an agreement with the provincial government and a logging company (with timber rights in the region) to protect the watershed.

Hundreds of miles to the south, in Oregon, a coalition of community residents have formed a partnership to save the Applegate watershed. Encompassing approximately 496,500 acres (about 200,000 hectares), the Applegate Watershed includes all the land around the water systems that make their way to the Applegate River. In the past, farmers, loggers, and environmentalists in this region frequently found themselves in conflict, hurling insults at each other. Once a pristine valley, the watershed had now been scarred by a maze of clearcuts, logging roads and sluggish brown streams. The Applegate Partnership brought together farmers, environmentalists, ranchers, loggers, educators, other residents, and officials from natural resource agencies to work on a long-term plan to restore the health of the watershed. Learning to work together, their motto became

"Practice Trust — 'them' is 'us'." Over time, the Applegate Partnership worked with government agencies to develop a "forest-health" plan to help restore the watershed, and then negotiated with logging companies. These and similar restoration plans for farm and ranch lands in the community have gradually brought the watershed back to life again.

Preservation of the local watershed was also the common theme that united people of a timber town tucked away in the northwest corner of California. During the 1970s and 1980s, the town of Hayfork, California, was savaged by timber wars that pitted loggers against environmentalists. After a court decision in 1990, which virtually shut down the logging industry in the area to protect the habitat of the endangered spotted owl, community residents of Hayfork began to realize that they would have to rebuild their local economy on a more sustainable basis if the community was going to survive. In 1992, a group of community stakeholders (including environmentalists, loggers, sawmill workers, restaurant owners, and officials from government agencies) began to meet around the common goal of restoring the local watershed. They set up a Watershed Research and Training Center and began to train themselves for the restoration of the area drained by the South Fork of the Trinity River. In 1993–94, assisted by the federal forest service, the group went through a process to develop a plan for a twenty-two-thousand-acre (about nine-thousand-hectare) watershed. Although the plan involved no timber sales, a series of projects were undertaken to revitalize the forests and waterways in the region.

On the other side of the United States, in the state of Maine, a river was set free in July 1999, after being trapped by a hydroelectric dam for 162 years. Since its construction in 1837, the Edwards Dam had damaged the Kennebec River's ecosystem and blocked the passage of fish. Although the dam owners had refused to build a fishway to allow migrants to make their way upstream to spawn, remnant populations had managed to survive below the dam year in and year out. Throughout the 1990s, a collection of conservation groups, calling themselves the Kennebec Coalition, waged a relentless lobbying and education campaign to have the dam decommissioned. Then, in 1997, the Federal Energy Regulatory Commission took the unprecedented step of ordering the removal of the dam, against the owners' wishes. Based on the coalition's campaign, including a seven-thousand-page brief, the commission had come to the conclusion that the economic and environmental benefits of a free-flowing Kennebec River were greater than the continued operation of the hydro dam. To celebrate the occasion, people wore T-shirts in the streets with the inscription: "River reborn. Kennebec flows free."

Restoring watersheds is by no means an activity exclusive to North American communities. To varying degrees, these kinds of fightback initiatives are taking place in every continent on the planet. In India, for example, anti-dam activists have been increasingly focusing attention on ways to provide irrigation and drinking water that do not rely on large dams. More specifically, civil society organizations are working with local communities to reinvigorate traditional water-harvesting systems. In South Africa, emphasis has been put on organizing "catchment communities," or communities within the rainfall catchment area (watershed) of a water system. Examples of catchment communities in South Africa include the Okavango Liaison Group, a regional coalition of local community groups committed to revitalizing the Okavango River and Delta, and the Greater Edendale Environmental Network (GREEN), a grassroots group working with a variety of organizations in the Pietermaritzburg-Msunduzi area of South Africa to create "a clean and safe Msunduzi River by the year 2009."

STOPPING DAMS

The anti-dam movement has been at the forefront of the fightback for water rights around the world. Over the past hundred years, says Patrick McCully of the mega-dam watchdog group International Rivers Network, some forty thousand large dams were built on the world's rivers, flooding nearly one percent of the earth's land surface and displacing up to 60 million people who, for the most part, became poorer as a result. At the same time, these dams caused untold destruction to ecosystems and biodiversity. But in 1981, an uprising by Indigenous peoples in the Philippines resulted in the cancellation of a World Bank-funded plan to construct a dam on the Chico River, thereby igniting a people's movement to save rivers and riverine communities all over the world.

Through actions by groups large and small, says McCully, this people's movement has grown to include thousands of environmental and human rights groups all over the world. As these groups gained a stronger voice, dams became less accepted as a way of providing power and water. The cost overruns of these mega-projects often resulted in an economic burden for the countries that paid for their construction, and the dams failed to deliver what was promised — an abundance of energy and water at cheap rates. But the protest movement has also made it increasingly difficult to construct huge dams on river systems in most countries. As McCully points out, dam construction peaked at 540 a year in the 1970s but decreased to 200 a year in the 1990s. Indeed, by 1992, the president of the International Commission on Large Dams, Wolfgang Pircher, was warning the industry that "a serious counter-movement . . . has already succeeded in reducing the prestige of dam engineering in the public eye, and it is starting to make work difficult for our profession."

Although there are many moving stories to be told about the fightback against mega-dams around the world, a few can help us grasp the significance of the movement. One series of stories comes out of the countries formerly in the Soviet Eastern Bloc. In Hungary, an independent citizens' group called the Duna Kor, or the Danube Circle, was illegally formed in the early 1980s to stop construction on the Nagymaros Dam. Given the authoritarian rule of the Hungarian Communist regime, Duna Kor's first objective was to break the secrecy surrounding the Nagymaros Dam project by circulating a petition calling for a parliamentary debate on the issue. And in 1985, Duna Kor published an environmental impact study on the project. But a year later, when the group held a press conference, activists were promptly arrested and interrogated, and this sparked a protest march. Through these skirmishes, the movement spread. In October 1988, fifteen thousand Hungarians filled the streets of Budapest, protesting against the damming of the Danube River. In May 1989, the now non-Communist Hungarian government stopped work on the Nagymaros, and in October of that year, Parliament passed a resolution to abandon the project altogether.

A similar struggle in Guatemala during the early 1980s ended in a massacre. In 1982, the Chixoy hydro dam in Guatemala, jointly funded by the World Bank and the Inter-American Development Bank, was nearing completion. When it came time to fill the reservoir, the Maya Achi residents in the riverside village of Rio Negro refused to leave their homes and their land. The resettlement site provided by the Guatemalan power utility offered cramped houses and poor land. In response, government-backed paramilitary forces carried out four massacres during an eight-month period in 1982, murdering 440 Maya Achi people in Rio Negro. "They killed us just for claiming our rights to our land," recalls Cristobal Osorio, who lost his wife and infant child, along with 19 other family members, in the massacre. And of course, the Rio Negro ecosystem was another casualty. Today, Osorio presides over a committee of 150 Rio Negro families who lost their loved ones, as well as their ancestral lands, to the Chixoy Dam. A UN Truth Commission has since denounced the atrocities as genocide and the World Commission on Dams has called for reparations to help rectify past wrongs.

More recently, communities affected by the construction of the Pak Mun Dam in Thailand have continued their protests since the World Bank-funded project was completed in 1994. Focusing their demands on both the World Bank and the Thai government, Pak Mun villagers are calling for the removal of the dam, the restoration of the river it blocked, and the recovery of the fisheries. (The drastic reductions in fish populations resulting from the construction of the Pak Mun Dam have directly affected the livelihoods of more than twenty-five thousand people.) In its report, the World Commission on Dams charged that the project had failed to deliver promised benefits and the fisheries had been seriously damaged as well. With this assessment, the Pak Mun communities felt they had a right to demand reparations for the losses they had suffered. But when Bank officials refused to acknowledge the project's deficiencies, let alone do anything about them, the villagers intensified their resistance. In March 1999, some five thousand villagers set up a "protest village" near the dam site. In November of that year, however, the protest village was raided, people were forcibly evicted, and fires were lit to burn the makeshift wooden shelters and destroy their campsite. Undaunted, the villagers vowed to continue their fight.

In the United States, members of the anti-dam movement have experienced a remarkable turn of events. Recognizing that the dam-building era is virtually over in that country, government officials have recently been focusing their attention on the decommissioning of dams, thereby freeing up natural river systems. In 1998, the head of the U.S. Department of the Interior set out on what was called a "sledgehammer tour" across the country, to remove a series of small, aging dams that were having negative impacts on local fisheries. By 2000, there were active campaigns in place working for the decommissioning of more than one hundred dams across the country. In carrying out these decommissioning campaigns, unique partnerships have been formed between river protection groups, fisheries experts, concerned citizens, local politicians, and scientists from various U.S. agencies. In Wisconsin, the "20 by 2000" campaign launched by the River Alliance hoped to have 20 dams in six communities removed or slated for removal by 2000. Friends of the Earth have also joined with American Rivers and local Indigenous peoples in an effort to bring about the decommissioning of dams in the Columbia-Snake River basin. And the Sierra Club is spearheading efforts to encourage the decommissioning of dams that submerged the Hetch Hetchy Valley in Yosemite National Park and the restoration of the rivers that ran through it.

Indeed, the anti-dam movement has reached a certain level of maturity. The accent is no longer simply on stopping the construction of mega-dams on major rivers. There is also a growing commitment to developing more sustainable, equitable, and efficient approaches to the management of rivers and to establishing a more democratic decision-making process that is rooted in the communities themselves. At the same time, there is a recognition among some anti-dam activists that more profound changes need to take place in the dominant economic and political system itself. In India, for example, the struggle against mega-dams in the Narmada Valley "has come to represent far more than the fight for one river," says renowned author Arundhati Roy. "It has begun to raise doubts about an entire political system. What is at issue now is the very nature of our democracy. Who owns this land? Who owns the rivers? its forests? its fish?"

INTERNATIONAL STRUGGLES

Although many citizen activists have chosen to focus their energies on local governments in their fightback campaigns on water issues, it is also clear that civil society groups in some countries are not prepared to let their national governments off the hook. In France, for example, citizen activists began to gear up for a major public debate over a parliamentary review of their national water legislation in 2001. In Canada, the federal government is being pressed to commit major financial resources to rebuilding the country's public infrastructure for water services and to develop a national policy banning water exports. And the campaigns being waged by water activists in Ghana and Uruguay are aimed at changing national government policies and legislation. In South Africa, the fightback in the townships over water cutoffs is largely aimed at ensuring that the national government adheres to the country's constitution, in which water is enshrined as a basic human right. Meanwhile, in the United States pressure from citizens helped lead to the decommissioning of river dams by federal agencies — a further indication that citizen activists have not entirely given up on their national governments as vehicles for democratic social change — even in this era of corporate-driven globalization.

At the same time, citizens' campaigns on water issues are becoming increasingly internationalized. Anti-dam battles, like those in the Narmada Valley of India, have become the focal point of work by citizen activists around the world, not simply because of the role played by global institutions like the World Bank and the IMF, but also because of the organizing efforts of civil society groups like the International Rivers Network. Similarly, the fightback to overturn the privatization of water services in Cochabamba, Bolivia, became internationalized not only because of the Bank and global corporate players like Bechtel, but also as a result of the organizing initiatives of the Coordiadora and its allies — including organizations like Public Services International. In fact, most citizen activists realize that these water campaigns cannot be won at the level of the local community alone. The increasingly globalized nature of the water industry and the market itself require that community-based campaigns take on international dimensions, in order to be effective in the long run. This is especially true when global water giants become the focal point of campaign struggles.

In addition, events like the World Water Forum held in The Hague in March 2000 have emerged as important international arenas for action.

Had it not been for the initiative of the Blue Planet Project, launched by the Council of Canadians in the year 2000, and allied groups like Public Services International, the vast majority of the delegates from countries around the world would have been exposed only to the dominant message — and sadly incomplete picture — provided by the global water corporations. Moreover, the fight for water rights is gaining a prominent place in anti-globalization campaigns being waged against what social movement activists in India call the "Unholy Trinity" of the IMF, the World Bank, and the World Trade Organization. For many years, anti-dam struggles have been a major flashpoint in international battles against the IMF and the Bank, but mass protests mobilized since 1999 have also focused on the way these global institutions are using their financial clout to force governments to privatize water services. Similarly, the fight for water rights has become one of the major issues in the international campaign being organized to stop or change the direction of the negotiations regarding the General Agreement on Trade in Services (GATS) that are currently taking place at the World Trade Organization.

This brief survey of people's fightback struggles for water rights around the world reveals that the seeds of resistance have not only been planted, but they are growing and multiplying. Yet there are also serious gaps and limitations. Clearly, the various kinds of people's movements that have been organized around water struggles so far do not begin to measure up to the scope of the global crisis at hand, let alone the power of the economic and political elites that are so energetically promoting their nonsustainable "solution." This should not be taken, however, as reason for discouragement. On the contrary, we are only in the early stages of building a global movement on water issues. Yet there are steps that can be taken now to increase the momentum of the world's water-preservation initiatives, unifying them around a set of common principles. Indeed, the momentum has already begun to build as the water-security movement joins forces with other environmental and social justice groups working to put a stop to the corporate takeover of the world's water supplies.

THE STANDPOINT

How common principles and
goals can save the world's water

University of Toronto Professor Emerita Ursula Franklin says that the most important social movements in history are grounded by the presence of a "standpoint." A standpoint, explains Dr. Franklin, is an ethical framework that informs one's purpose and one's work. "Where you stand and where you look tell you what is in the foreground, what is in the background, what is big and what is small." A standpoint brings a sense of priority, a sense of proportion, and a sense of obligation. Having the courage to find a place to stand, and if necessary, fight for what you believe, is required before any person or movement can effect real social change. The tragedy of most modern governments, says Franklin, is that they have embraced economic globalization, which denies the standpoint of community or environmental stewardship in favor of the sole standpoint of profit. For governments and corporations, the profit is big and in the foreground, and the care for people, Nature, and democratic principles has vanished. If the world's water is to be saved for future generations, millions of the world's citizens will have to take a stand based on a set of principles and ethical considerations directly opposed to the predominant standpoint of the global economy.

At the dawn of the millennium, the world is poised to make crucial, perhaps irrevocable, decisions about water. Individuals, countries, and corporations all over the globe are still polluting the very water that gives them life. Treatment systems are nonexistent or pushed to the limit in the poorer countries, and even in the wealthier nations, hormones and deadly chemicals are found in local water supplies. Some polluters keep pouring poisons into water systems even when confronted with evidence of the damage they have created, but the harm done to water to date has been largely unintentional and reactive — a combination of benign neglect, ignorance, greed, too many demands on a limited resource, careless pollution, and reckless diversion. On the whole, the human race has taken water for granted and massively misjudged the capacity of the earth's water systems to recover from our carelessness. Although we must now answer to the great harm we have caused, it is probably fair to say that no one set out to create a global water shortage or to deliberately destroy the world's water supply.

CROSSROADS

However, lack of malice is no longer a good enough excuse. We know too much. We know how careless environmental practices such as clearcutting and toxic dumping are destroying waterways. We understand the connection between energy-hungry industrial and personal practices and the global warming which is destroying aquatic habitat. There is mounting evidence that we are depleting aquifers at a totally unsustainable rate but we keep on mining groundwater supplies because we won't stop polluting surface water. And we know that our irrigation practices are not only leading to desertification of land, but destroying water tables as well.

Yet societies all over the world, or at least their governments and private sector leaders, have bought into tenets of economic globalization based on a model of unlimited growth and the allure of unchecked consumerism. We continue to create conditions that force small farmers to abandon their land and head for overcrowded cities. We implement global trade policies that reward ecologically unsustainable production methods of goods and food. We favor governments that generate low consumer prices by cutting back on domestic regulation of agriculture, food production, chemical use, and industrial dumping. In fact, almost everything we do in modern industrialized society is guaranteed to deepen the global fresh water crisis. Huge transnational corporations are operating under the umbrella of trade regimes like the North American Free Trade Agreement, which make governments roll back environmental legislation for fear of reprisals at trade tribunals. Meanwhile, corporations pay minimal taxes or hide their profits in tax havens, thus diminishing potential government revenue that could have paid for water infrastructure improvements, sanitation services, and safe water practices. Our leaders have entrusted our lives to those driven solely by the profit imperative.

And now, as we have seen, transnational corporations, backed by international trade and financial institutions like the International Monetary Fund and the World Bank are moving in to profit directly from the global fresh water crisis. If we allow these private sector companies to take control of the world's water supplies, we will lose the capacity to save the world's water. We will be allowing the emergence of a water elite that will determine water use based on its own interests. Instead, we should be working to help people and communities around the world take responsibility for a shared resource and treat it in ways that will ensure a water-secure future for their descendants.

The move to commodify depleting global water supplies is wrong — ethically, environmentally, and socially. It ensures that decisions regarding the allocation of water center almost exclusively on commercial, not environmental or social, considerations. Corporate shareholders seek maximum profit, not sustainability or equal access. Privatization means that the management of water resources is based on principles of scarcity and profit maximization rather than long-term sustainability. Corporations are dependent on increased consumption to generate profits and are therefore much more likely to invest in the use of chemical technology, desalination, and water diversion than in conservation.

And the trend to commodify what has been a public service makes it much more difficult for citizens to allocate and manage their own water sources. The concentration of power in the hands of a single corporation and the inability of governments to reclaim management of water services once a private water supplier has been contracted allows corporations to reduce the democratic power of citizens. And as transnational water companies lobby to reduce environmental regulations and deregulate water standards, they gain undue influence over government policies.

In spite of the obvious dangers it presents, the commodification of the world's fresh water is advancing at an alarming rate. Decision making over this precious resource appears to have fallen into a relatively small number of hands — bureaucrats at the World Bank and the UN, a cadre of professional water experts who advise them, government aid agencies, trade economists, and powerful water corporations with a personal stake in the outcome. This small but powerful group has determined that the debate is over; "everyone" supports the privatization of water, it asserts. This, of course, is patently untrue. The citizens of the world have not been consulted or even informed of this development. In fact, as outlined in Chapter 8, evidence is mounting that when citizens are given the choice about the control of their water, they opt for public, local, and transparent control.

What seems clear is that governments are not going to take the lead in this debate; instead, citizens will be forced to create the political arguments that will keep the world's water in the public "commons" for all time. To do so, general agreement must be reached about the fundamental principles needed to ensure a water-secure future for the world — and to come to agreement on these principles, five ethical questions about water must be addressed, questions concerning the commons, stewardship, equality, universality, and peace.

THE WATER COMMONS

The antidote to the commodification of water is its decommodification. Water must be declared and understood for all time to be common property. In a world where everything is being privatized, citizens must establish clear perimeters around those areas that are sacred to life or necessary for social and economic justice. Equal access to water is absolutely central to both life and justice.

As Indian physicist and activist Vandana Shiva points out, commonly owned water is not destroyed — as long as its use is controlled by conservationist rules. In fact, the only strategy for conserving water that has proved successful in time of scarcity is the renewal and rejuvenation of common property rights so that patterns of use are governed by Nature's limits of renewability and the social limits of water equity. Those who hold water in common must set rules as to its use because a few individuals within a group could take more than their fair share if there were no disincentives. "Privatizing water through property rights will not reverse this degradation," says Shiva, "it will accelerate it. It will unleash water wars by pitting person against person, region against region, the rural areas against privileged and rich urban centers and the poor against the rich."

International water crusader Riccardo Petrella explains that it is an essential feature of the market that one should be able to choose among several goods of the same or a different nature, using for that choice such criteria as price and quality. The argument for commodifying water is the same argument used for the commodification of widgets: the marketplace is the optimum model for the most efficient allocation of material and natural resources, as well as the distribution of wealth. That is, each country will produce what it does best, and all will compete in the open marketplace. Hence, wealthy countries market technology, ideas, and telecommunications, while poor countries, with cheap labor, export goods made under poor conditions. Countries that are wealthy in natural resources like oil or water also "compete," selling these "products" in the global marketplace. According to this argument, government standards and subsidies for export are merely impediments to "efficient" competition in an open market.

However, to have access to water is not a matter of choice or of efficient accumulation of wealth; it is a matter of life and death. Water is not something to be bought and sold for profit like a pair of shoes or a pizza. It's true that water-bottling companies are marketing their "goods" just like hats or gloves or cars, presenting them in an amazing array of "options." But all of this is an illusion, of course. In addition, water is too precious a resource to be processed and distributed according to profit principles, which unleash a juggernaut of ever-accelerating consumption and ever-expanding markets. All bottled water comes from the same finite source. Supplies cannot be infinitely increased to serve continually burgeoning markets. There is no source of life comparable to water within the eco-system, apart from soil and air. Water is unique, of limited supply, irreplaceable, and necessary for all life. The very fact that it cannot be replaced with anything else makes water a basic asset that cannot be subordinated to the principles of the marketplace. Thus, says Petrella, water is essential to the functioning of society as a whole and is therefore a social asset and a common good basic to any human community.

Vandana Shiva adds that water markets will not guarantee water availability to all, but will only guarantee access to the economically powerful while excluding the poor and the marginalized. As we become slaves to the whimsical dynamics of the deregulated "free" market, the commons is being destroyed and the weaker sections of society are being denied their right of access to a resource that is essential to their health and life.

This tragic outcome is completely unnecessary. Instead of commodifying water even further, we need to recover it by treating it as part of the commons and by strengthening community participation in water management — according to conservationist principles. Activist groups in South Africa have pointed out that when water is treated as part of the commons — a right to which all individuals are entitled — it is supplied to more people, on a more equitable basis, than if it was subjected to the dynamics unleashed by the profit motive. This means that more people have better health and therefore have greater ability to contribute to the common good. This generates economic activity. At the same time, the resource is conserved because the profit motive has not driven suppliers to produce more and more until their water sources dry up. This enhances the health of the earth and helps maintain the balance of its ecological processes. And when the planet is healthy, it can better support responsible, sustainable economic activity, thus enhancing the prosperity of its citizens. In other words, the water commons not only recognizes the right of the individual to water for life, but also promotes the common good. As Petrella argues, this is why every society must collectively cover the costs of providing basic access to water for all. It is a basic human obligation and it makes long-term ecological and economic sense.

WATER STEWARDSHIP

It is becoming distressingly clear that in most modern industrialized societies, humanity has lost touch with the natural world in a way that threatens the very existence of the planet. Instead of respecting the way Nature has distributed water, we have massively attempted to tame, alter, and control water systems to suit our needs. The results have been catastrophic. Such behavior comes from a worldview that places humanity above Nature and God and has permitted us to live outside Nature's laws for some time. But clearly, this irreverence is now striking back at us with a vengeance.

At the heart of any new water ethic must be a renewal of our ties with the natural world and a reverence for water's sacred place in it. Humans must see ourselves as one species among many, whose existence, like that of all species, depends on our living within the rules of the natural world. Instead, humans have polluted their own common property — a *modus operandi* that is anathema to the more reverent and logical way of using resources that must define our future. If we are to survive as a species, our lakes, rivers, streams, and groundwater supplies must be preserved and reclaimed, and all economic and human activity must conform to this goal.

This new water ethic includes a radical rethinking of the issues surrounding water diversion, large dams, and large-scale irrigation projects. Our water future must encompass a radically altered approach to water systems based on more sustainable, equitable, and efficient technologies and more earth-friendly agricultural practices. Large corporate farms and the technologies and chemicals that sustain them must go. Importantly, we must resist the call to construct massive new technological systems to move bulk water around the globe by tanker, pipeline, canal, and river diversion.

Those who view water as a commodity say that water flowing into the sea or situated in what one forest company CEO calls "decadent wilderness," is not of service to people or the economy and is therefore a wasted commodity. At the root of this viewpoint is the assumption that all resources must be captured and made available for sale, in order to increase monetary wealth. The main error of this argument is that resources need to replenish themselves, and water flowing into the ocean is part of a natural hydrological cycle that has balanced the earth's ecosystems for millennia. To assume that these processes can be interrupted and distorted permanently on a massive scale is to tamper with systems that have sustained life on earth with great success. By moving large volumes of water from one area of the globe to another, we will disrupt natural breeding grounds of birds, amphibians, and mammals, and we will be creating new drought zones as aquifers are depleted and rivers and lakes are dried up through damming and diversion.

Scientists also warn that removing vast amounts of water from watersheds has the potential to destroy ecosystems. Lowering water tables, for instance, can create sinkholes and dry up wells. Huge energy costs would also be associated with large-scale water movement. In Canada, one version of the hypothetical GRAND Canal scheme (to divert water from rivers flowing into James Bay) called for a series of nuclear power stations along the route to supply the energy needed for the movement of such huge volumes of water. And existing water diversions and hydroelectric megaprojects are already causing local climate change, reduced biodiversity, mercury poisoning, loss of forest, and the destruction of fisheries habitat and wetlands. Yet the negative ecological impacts of all these projects pale in comparison to those that will transpire if massive water-displacing technologies are implemented in the future.

Scientific studies show that large-scale water removal affects more than just immediate ecosystems. Commenting on the importance of maintaining natural river flows, in order not to imbalance seacoast ecosystems, Canadian water expert Jamie Linton points out that "beyond all doubt, water is not 'wasted' by running into the sea . . . [and] the cumulative effects of removing water from lakes, rivers and streams for export by tanker could have large-scale impacts on the coastal and marine environment." Similarly, Canadian writer and filmmaker Richard Bocking, who specializes in water issues, says we strike a Faustian bargain when diverting rivers. "For power generation or irrigation today, we exchange much of the life of a river, its valley and biological systems, and the way of life of people who are in the way. As the cost of the last 50 years of dam building becomes evident, we can no longer plead that we don't know the consequences of treating rivers and lakes as plumbing systems."

Depending on desalination technology is also a Faustian bargain. While desalination projects do currently serve some communities and countries, and while the use of this technology is likely to grow in the future, desalination is not the panacea for the world's water crisis. It is prohibitively expensive, so at least for the foreseeable future, it is available only to wealthy countries. Even if costs fall, however, desalination is highly energy intensive; it cannot be done without injections of massive amounts of fossil fuels, which would just add to global warming — already an enemy of the world's fresh water supplies.

Furthermore, desalination also produces a lethal by-product. For every gallon of sea water processed, only one-third becomes fresh water. The remaining two-thirds is a highly saline brine that, when dumped back into the ocean at high temperatures, is a major source of marine pollution. And desalination of some seawater doesn't begin to touch the problem of the growing salination of groundwater supplies. Common sense must tell us that it would be easier to stop current practices that cause the salination of existing fresh water supplies than to set up expensive, climate-change-enhancing practices to desalinate the world's oceans. Quite simply, the misuse of major intrusive technologies has been a major factor in getting us into this mess in the first place. Adding more of the same is not the answer to our fresh water crisis.

WATER EQUALITY

What, however, of the humanitarian argument that in a world of water inequality, water-rich areas have an obligation to share water supplies with others? Any comprehensive and sustainable water ethic would of course acknowledge this fact. Those who are water-poor live almost exclusively in the Third World; those who are water-rich live in the industrialized nations, where corporations and certain classes of people have become wealthy from the colonization of many of the areas now living in water-stressed conditions. This represents a tragic dilemma. It could be argued that the industrialized nations have a moral obligation to share with water-poor areas, even though this would put great stress on already damaged ecosystems. To try to resolve the dilemma, it would be helpful to distinguish between short-term and long-term approaches. Importing water is not a desirable long-term solution for either the ecosystems or the peoples of water-scarce regions of the world. Since water is essential to life, no one should become dependent on foreign supplies that could be cut off for political or environmental reasons. It would also be helpful to distinguish between water trading and water sharing. In a commercially traded water exchange, those who really need the water would be the least likely to receive it. Water hauled long distances by corporate-owned tankers would be available only to the wealthy, especially large corporations, because the motive of export is ever-increasing profit. Importing water in this way for only those who could afford it would reduce the urgency and political pressure to find real, sustainable, and equitable solutions to water problems in water-scarce countries.

George Wurmitzer, the mayor of Simitz, a small town in the Austrian Alps, captures the difference between water sharing and water trading when he expresses concerns about large-scale exports of water from his community: "From my point of view, it is a sacred duty to help someone who is suffering from thirst. However, it is a sin to transfer water just so that people can flush their toilets and wash their cars in dry areas. . . . It makes no sense and is ecological and economic madness." And as water expert Jamie Linton says, "Perhaps the strongest argument against [commercial] water export is that it would only perpetuate the basic problem that has caused the 'water crisis' in the first place — the presumption that people's growing demands for water can and should always be met by furnishing an increase in the supply. This thinking has led to the draining of lakes, the depletion of aquifers and destruction of aquatic ecosystems around the world."

If, however, we maintain public control of water, it might be possible to share water supplies on a short-term basis between countries in times of crisis. In these cases, water sharing would need to be accompanied by strict timetables and conditions aimed at making the receiving region water independent as soon as possible. This way, water could be used to encourage water system restoration. This kind of resolution is not conceivable, however, if the privatization of the world's water supplies and distribution continues unchallenged; corporations would not allow a nonprofit system of water transfer to be established.

~

At the root of the global water inequity crisis is the deep and growing gap between the nations of the North and those of the South. Most Third World countries have little or no capacity to provide sanitation services to their populations or to halt the spread of water-borne diseases. And in their inability to set conditions on foreign investment, they have little power to force industry to stop polluting their waterways. Moreover, the IMF and the World Bank have imposed food production and export practices that promote nonsustainable, water-guzzling farming. These policies must end, along with those imposed by the North forcing Third World governments to abandon public health services to the poor — including the delivery of public water.

To truly deal with the global water poverty gap, wealthy nations must share their financial resources not to promote nonsustainable water systems whose main purpose is to garner huge profits for large transnational corporations, but to establish sustainable water systems. There are several immediate actions governments could take. These include canceling Third World debt; bringing foreign aid budgets back up to their previous standards (0.7 percent of GDP); and implementing a "Tobin Tax" on financial speculation that would pay for water infrastructure and universal water services.

Further, the special plight of Indigenous peoples around the world must be recognized and given special attention. In many countries, First Nations' water rights have been expropriated for profit. Indigenous peoples have been disproportionately hurt by the construction of megadams and water diversion projects, and their lands and waters have been polluted by industry to an exceptional extent. Water is also a foundation of spiritual life for Indigenous peoples — a further reason that their proprietary interest in waters on their traditional lands must be respected and honored.

WATER UNIVERSALITY

Linked to the question of water equality is the issue of pricing and how it affects equitable and universal access to water sources. In many circles around the world, there is a growing call to give water true economic value by pricing it on a cost-recovery basis. Increasingly, environmentalists and others rightly point out that in many water-rich countries, water is taken for granted and badly wasted. The argument is this: if an economic value were to be put on water, people would be more likely to conserve it. This idea might appear to be a good method of conservation, but in the current climate, there are serious concerns that need to be raised about the issue of water pricing.

First, water pricing exacerbates the existing global inequality of access to water. As we know, the countries that are suffering the most severe water shortages are home to the poorest people on earth. To charge them for already scarce supplies is to guarantee growing disparities between the rich and the poor and their ability to obtain water. As a result, water pricing will also widen the North-South divide. Already, because many in the North blame the South's population explosion for the global water shortage, the call to "just price it" may in some cases be a thinly veiled argument to let high water prices curb population. One hears echoes of this thinking from those who claim that HIV/AIDS is "Nature's" answer to overpopulation in the Third World.

The privatization of this scarce resource will lead to a two-tiered system within the industrialized world as well — those who can afford water and those who cannot. It will force millions to choose between necessities such as water and health care. In England, for instance, under the Thatcher regime, high water rates forced people to choose whether or not to wash their food, flush their toilets, or even bathe.

Second, under the current rules of trade agreements and financial institutions like the World Bank, priced water is considered a private commodity. Only if water is maintained as a public service, delivered and protected by governments, can it be exempted from the onerous enforcement measurements of these powerful bodies. The trade agreements, like the WTO and NAFTA, for instance, are very clear: if water is privatized and put on the open market for sale, it will go to those who can afford it, and that does not necessarily mean those who need it most desperately. Once the tap has been turned on, by the terms of trade rules, it cannot be turned off. Further, under free trade agreement rules, governments cannot set a two-price regime. The market would determine the cost of water, and governments, even those exporting it, would have to charge their own citizens the higher prices commanded and set on the global market. Poor people in rich and poor countries alike would suffer.

The World Bank has appeared to show concern for the poor who will be cut off from the water supplies that will become unaffordable after privatization. They have encouraged poor nations to subsidize water for their people. Chile, for instance, has introduced "water stamps" for the country's poorest residents. But anyone familiar with the problems of welfare, particularly in the Third World, knows that such charity is often nonexistent and punitive where available. (Governments, for instance, will often provide free water to communities that vote for the ruling party.) Water as a fundamental human right is guaranteed in the UN's International Covenant on Economic, Social and Cultural Rights. Water welfare is not what the architects of that great declaration had in mind.

Third, as it is now envisaged, water pricing won't do too much to conserve water. It is generally accepted that water consumption in urban centers breaks down at 65 to 70 percent industrial, 20 to 25 percent institutional, and 10 percent domestic. Yet most of the discussions about water pricing center on individual water use, while large corporate users consume vast quantities and some have been notorious for avoiding the cost of their water altogether.

Finally, in an open bidding system for water, who will buy it for the environment and the future? In all of this privatization/pricing debate, there is precious little said about the natural world and other species. That is because the full cost of water extraction and distribution, including the cost to the environment, has not been factored into the commercial equation. And this cost must be taken into account and legislation must be brought forward that reflects that accounting. If we lose public control of our water systems, we will have no legislative power to protect the vital watersheds and pristine lakes from misuse and possible destruction.

The dialogue about water pricing is a crucial one; however, it must take place within a larger framework. To be both effective and just, any serious consideration of water pricing must take into account three factors: the global poverty gap; water as a human right; and water in Nature. If water is sold, it must be through a fair pricing system based on ability to pay, guaranteed free water for basic necessities, and a just tax system. Then, the revenues raised from this fair pricing system must be made available to remedy water problems in any part of the world and to work toward ensuring universal access to water. Specifically, monies raised through fair pricing would help make the following improvements: ensure that a basic amount of water and basic sanitation are guaranteed to every person on earth regardless of ability to pay; protect the environment and restore watersheds; enforce clean water standards; and repair faulty infrastructure, which is currently the cause of great amounts of water wastage.

To attain universality of access to water, governments must also focus on passing and enforcing laws to stop wastage by the greatest abusers of water — big industrial enterprises and large, agribusiness farming operations. Governments must also implement a more just taxation system that captures some of the untold billions in tax dollars that large corporations currently evade through the use of tax havens and by convincing governments to lower or abolish taxes to attract industry. These revenues would go a long way to cleaning up the earth's dying water systems. Clearly, the focus must be on those who use water most and who thereby remove the benefits of this common good from the community in the form of profits. In an age of mergers and transnational operations, common goods or public trusts like water can be moved thousands of miles away from their community of origin. Business, however, has no right to deprive anyone of inalienable human rights such as access to clean water; if that is the price of profit, the price is too high.

Implementation of these methods would go a long way to ensuring universal access to water and good sanitation. But no implementation will be possible if water is not controlled in the public interest. If, instead, governments allow corporations to control and commercialize water, the profit principle will dominate. In this case, water pricing would become a tool of the market, rather than an incentive for conservation and justice. Since water is essential for life, universal access to water is a basic human right, and this right must be a basic principle of a new water ethic.

WATER PEACE

In a world increasingly plagued by water shortages, the number of cross-border conflicts is bound to rise unless humans realize that the need to address this common threat is greater than any differences among us. In fact, unless the human race takes immediate measures to collectively confront the global water crisis, the future for a water-secure world is very bleak indeed. Somehow, we must come to a common understanding of the nature and size of the crisis in order to take the collective actions that are needed. If, for instance, scientists could prove that a huge comet with a capacity to destroy the world was hurtling toward earth and that only the concerted collective action of all humanity might deflect it, can we doubt that race, religion, and ethnic and socio-economic differences would suddenly not matter so much? Yet a global water catastrophe is hurtling toward us, and in the end, no one will be spared.

The international community will have to come together under a new model — one that is not driven by the profit motive or funded by the World Bank. It will have to embrace the values of water commons, water stewardship, water equality, water universality, and water peace. Some positive precedents that support these values already exist — including three principles that have been developed over the last decade at international meetings sponsored by the United Nations. These are the principles that could form the basis of the way we govern water use in the future:

* *the principle of limited and integrated territorial sovereignty*, according to which every state has a right to use the waters in its territory on condition that this does not harm the interests of the other states;

- *the principle of a community of interests*, according to which no state may use the waters in its territory without consulting other states to achieve integrated management based upon cooperation; and
- *the principle of fair and reasonable use*, according to which each state has a right to use the waters of the shared basin by being awarded ownership and control of a fair and reasonable share of the basin's resources.

As helpful as these first principles are, there are a number of concerns that they do not adequately address. First, most international or binational agreements on water conflicts are about disputed *shared-water* systems — usually rivers or lakes and sometimes aquifers. Increasingly, however, citizens within countries are running out of water and are demanding some access to the water that is *inside the borders* of other countries. This is a new development and needs its own principles.

Second, agreements based upon these principles are made between nation-states and are based on the kind of agreements that only nation-states can settle. But increasingly, the private sector controls water rights and water distribution in many countries. The move to commercialize water replaces nation-state authority with corporate authority, and this can place decision making about water beyond the scope or jurisdiction of sovereign nations.

Third, using their superior powers and enforcement mechanisms, trade and financial institutions like the WTO and the World Bank have the capacity to override international government decisions about water, including decisions made by the UN. Until some new institutions with rival powers are created, international agreements on water and the environment or water and social justice can be overridden by the interests of global trade and investment.

In the end, humanity will find itself up to the task or not. If not, many millions will perish and eventually, perhaps the earth itself will become uninhabitable. But if we are able to see not only the nature of the threat but also the possibility of greater global harmony that could come from such a quest, we could perhaps, through the search for fresh water, approach the elusive global peace that many have felt is possible.

TEN PRINCIPLES

To save our diminishing water resources and stave off further conflict, all levels of government and communities around the world need to begin working together, as they have done in the past to rebuild communities in the aftermath of war. But to embark on this mission, we must soon come to agreement on a set of guiding principles and values. We have outlined ten here, as a starting point for dialogue and action that will lead to renewal.

1. Water belongs to the earth and to all species.
2. Water should be left where it is whenever possible.
3. Water must be conserved for all time.
4. Polluted water must be reclaimed.
5. Water is best protected in natural watersheds.
6. Water is a public trust, to be guarded by all levels of government.
7. Access to an adequate supply of clean water is a basic human right.
8. The best advocates for water are local communities and citizens.
9. The public must participate as an equal partner with government to protect water.
10. Economic globalization policies are not water-sustainable.

1. Water belongs to the earth and to all species.

Without water, humans and other beings would die and the earth's systems would shut down. Unfortunately, however, modern society has lost its reverence for water's sacred place in the cycle of life as well as its centrality in the realm of the spirit, and this loss of reverence for water has allowed humans to abuse it. Over time, we have come to believe that humans are at the center of the universe, not Nature, and decision makers have forgotten to take into account the fact that water belongs to the earth, to all species, and to future generations. They have shunted these stakeholders aside and failed to consider their interests. For all our brilliance and accomplishment, we are a species of animal who needs water for the same reasons as other species. Unlike other species, however, only humans have the power to destroy the ecosystems upon which all depend, and that is the destructive path on which we have embarked.

Only by redefining our relationship to water and recognizing its essential and sacred place in Nature can we begin to rectify the wrongs we have done. Only by considering the full impacts of our decisions on the ecosystem can we ever hope to replenish depleted water systems and to protect those that are still unharmed.

2. Water should be left where it is whenever possible.

Nature put water where it belongs. Tampering with Nature by removing vast amounts of water from watersheds has the potential to destroy that ecosystem and systems far beyond it. The large-scale removal of water from lakes, rivers, and streams has disastrous impacts on the land surrounding them and on the coastal environments where rivers eventually empty into the sea. Diversion and destruction of healthy bodies of water also destroys the local economy of Indigenous peoples and others who depend on them for their livelihoods.

While there may be an obligation to share water (and food) in times of crisis, it is not a desirable long-term solution. When one country or region becomes dependent on another for its water supplies, it is in a precarious position. Modern transportation and technology have also blinded many of us to the fact that importing water from long distances is neither cost-effective nor secure. If the full environmental costs of dam building, water diversion, and tanker transport were taken into account, we would see that the globalized trade in water makes no sense. By importing this basic need, a relationship of dependency would be established that is good for neither side. Instead, we need to learn the nature of water's limits and to live within them, looking at our own regions, communities, and homes for ways to meet our needs, while respecting water's place in Nature. Then, in times of emergency, watersheds and groundwater sources will be in better condition and available to temporarily help those who are far away.

3. Water must be conserved for all time.

Each generation must ensure that the abundance and quality of water is not diminished as a result of its activities. This will mean radically changing our habits, particularly concerning water conservation. People living in the wealthy countries of the world must change their patterns of water consumption, especially those in water-rich bioregions. If we do not change these habits, any reluctance to share our water — even for sound environmental and ethical reasons — will rightly be called into question.

If we change these patterns, we may maintain sustainable groundwater supplies, ensuring that extractions do not exceed recharge. In addition, some water destined for cities and agribusiness will have to be restored to Nature and small and medium-sized farming operations. Government subsidies of wasteful corporate practices and industrial farming must therefore end. By refusing to subsidize such abusive water use and by rewarding water conservation, governments will send out the message that water is not abundant and cannot be wasted.

Large tracts of aquatic systems must also be set aside for preservation, and governments must agree on a global target. Planned major dams and river diversions must be put on hold while better solutions are found, and some existing river diversions need to be reoriented back to their natural seasonal flow. Governments everywhere need to put priority on improving aging and broken water-delivery infrastructure. Huge amounts of water are lost every year because of leakage, and aging pipes can carry disease-bearing organisms.

4. Polluted water must be reclaimed.
The human race has collectively polluted the world's water supply and must collectively take responsibility for restoring it. Water scarcity and pollution are caused by economic values that encourage the overconsumption and inefficient use of water, which are already depleting aquifers and will eventually put our health and lives at risk. A resolution to reclaim polluted water is an act of self-preservation. Our survival, and the survival of all species, depends on restoring naturally functioning ecosystems.

Governments at all levels and communities in every country must clean up polluted water systems and stop the unthinking and rampant destruction of wetlands and watersheds. More rigorous laws must be passed and enforced to control water pollution from agriculture, municipal discharge, and industrial contaminants — the leading causes of water degradation.

Governments must regain control over transnational mining and forestry operations, whose unchecked practices continue to cause untold damage to water systems.

Furthermore, the water crisis cannot be viewed in isolation from other major environmental issues such as clearcutting of forests and human-induced climate change. The destruction of waterways due to clearcutting severely harms fish habitat. Climate change will cause (and is already causing) more extreme weather conditions: floods will be higher, storms will be more severe, and droughts will be more persistent — the pressure on existing fresh water supplies will be magnified. To restore so much damaged water, countries will have to make international commitments to dramatically reduce human impacts on climate.

5. *Water is best protected in natural watersheds.*

The future of a water-secure world is based on the need to live within naturally formed "bioregions," or watersheds. The surface and ground-water conditions in these watersheds act as a set of parameters that govern virtually all of life in that region, including flora and fauna, which are related to the area's hydrological conditions. Living within the ecological constraints of a region is called bioregionalism, and watersheds are an excellent starting point for establishing bioregional practices.

Another advantage of thinking in watershed terms is that water flow does not respect nation-state borders. Watershed management is therefore one way to break the gridlock among international, national, local, and tribal governments that has plagued water policy around the world for so long. Thinking in terms of watersheds, not political or bureaucratic boundaries, will lead to more collaborative protection and decision making.

6. *Water is a public trust, to be guarded by all levels of government.*

Because water, like air, belongs to the earth and to all species, no one has the right to appropriate it or profit from it at someone else's expense. It is a public trust that must be protected by all levels of government and by communities everywhere. This means that water should not be privatized, commodified, traded, or exported in bulk for commercial purposes. To ensure that this rampant commercialization does not take place, governments all over the world must take immediate action to declare that the waters in their territories are a public good and to enact legislation to protect them. Water should also be exempted from all existing and future international and bilateral trade and investment agreements, and governments must ban the commercial trade in large-scale water projects.

While it is true that governments have failed badly in protecting their water heritage, it is only through democratically controlled institutions that this situation can be rectified. If water becomes clearly established as a commodity to be controlled by the private sector, decisions about water will be made solely on a for-profit basis, and individual citizens will have no say as to how the resource will be used.

Each level of government must protect its water trust. At the municipal level, urban centers should no longer divert water resources from rural areas to service their own needs. At the municipal and regional levels, watershed cooperation should be carried out to protect larger river and lake systems. National and international legislation should apply the rule of law to transnational corporations and end abusive corporate practices. Governments should tax the private sector adequately to pay for infrastructure repair. And all levels of jurisdiction should work together to set targets for global aquatic wilderness preserves.

7. Access to an adequate supply of clean water is a basic human right.

Every person in the world has a right to clean water and healthy sanitation systems no matter where they live. This right is best protected by keeping water and sewage services in the public sector, regulating the protection of water supplies, and promoting the efficient use of water. This is the only way to preserve adequate supplies of clean water for people in water-scarce regions.

It is also vital to remember that Indigenous peoples have special inherent rights to their traditional territories, including water. These rights stem from their use and possession of the land and water in their territories and their ancient social and legal systems. The inalienable right of self-determination of Indigenous peoples must be recognized and codified by all governments, and water sovereignty is fundamental to the protection of those rights.

In addition, governments everywhere must implement a "local sources first" policy to protect the basic rights of all citizens to fresh water. Legislation that requires all countries, communities, and bioregions to protect local sources of water and seek alternative local sources before looking to other areas will go a long way to halting the environmentally destructive practice of moving water from one watershed basin to another. "Local sources first" must be accompanied by a principle of "local people and local, small-scale farmers first." Agribusiness and industry, particularly large transnational corporations, must fit into a "local-first" policy or be shut down.

This does not mean that water should be "free" or that everyone can help themselves to limitless quantities. However, a policy of water pricing that guarantees an essential amount of water to every human would help conserve water and preserve the rights of all to have access to it. Water pricing and "green taxes" (which raise government revenues while discouraging pollution and resource consumption) should place a heavier burden on agribusiness and industry than on private citizens, and funds collected from these sources should be used to provide basic water supplies for all.

8. The best advocates for water are local communities and citizens.

Local stewardship, not private business, expensive technology, or even government, is the best protector of water security. Only local citizens can understand the overall cumulative effect of privatization, pollution, and water removal and diversion on their own community. Only local citizens know the effect of job loss or loss of nearby farms when water sources are taken over by big business or diverted for far-away uses. Local citizens and communities are the front-line "keepers" of the rivers, lakes, and underground water systems upon which their lives and livelihoods rest. They need to be given the political power to exercise that stewardship effectively.

Reclamation projects that work are often inspired by environmental organizations and involve all levels of government and sometimes private donations. But in order to be affordable, sustainable, and equitable, the solutions to water stress and water scarcity must be locally inspired and community-based. If they are not guided by the common sense and lived experience of the local community, they will not be sustained.

In water-scarce regions, traditional Indigenous practices, such as local water sharing and rain catchment systems that have been abandoned for new technology, are already being revisited with some urgency. In some areas, local people have assumed complete responsibility for water distribution facilities and established funds to which water users must contribute. The funds are then used to provide water to all in the community. Techniques like this should be applied in other water-scarce regions of the world.

9. The public must participate as an equal partner with government to protect water.

A fundamental principle for a water-secure future is that the public must be consulted and engaged as an equal partner with governments in establishing water policy. For too long, governments and international economic institutions such as the World Bank, the OECD, and trade bureaucrats have been driven by corporate interests. Even in the rare instances that they are given a seat at the table, nongovernmental organizations (NGOs) and environmental groups are typically ignored. Corporations that heavily fund political campaigns are often given contracts for water resources — and too frequently, corporate lobby groups actually draft the wording of agreements and treaties that governments then adopt. This practice has created a crisis of legitimacy for governments everywhere.

Processes must be established whereby citizens, workers, and environmental representatives are treated as equal partners in the determination of water policy and recognized as the true inheritors and guardians of that irreplaceable resource.

10. Economic globalization policies are not water-sustainable.

The values of unlimited growth and ever-increasing international trade inherent in economic globalization are incompatible with the search for solutions to water scarcity. Designed to reward the strongest and most ruthless, economic globalization locks out the forces of local democracy so desperately needed for a water-secure future. If we accept the principle that to protect water we must attempt to live within our watersheds, the practice of viewing the world as one seamless consumer market must be abandoned.

Economic globalization undermines local communities by allowing for easy mobility of capital and the theft of local resources. In addition, liberalized trade and investment enables some countries to live beyond their ecological and water resource means while others abuse their limited water sources to grow crops for export. In wealthy countries, cities, agribusiness, and industries are mushrooming on deserts. A water-sustainable society would denounce these practices.

Global sustainability can be reached only if we seek greater regional self-sufficiency, not less. Building our economies on local watershed systems is the only way to integrate sound environmental policies with peoples' productive capacities and to protect our water at the same time.

Although world water supplies are dwindling and transnational corporations are working hard to reap substantial profits from that scarce supply, it is not too late to turn the situation around. Universal and equitable access to water is possible. The water commons can be saved from those who are already invading, to use it for their own profit. Private citizens do not have to stand by and watch as bottling companies move into their area, drain aquifers, fill their own pockets, and then depart. They don't have to put up with the privatization of water services. The people most affected by water-guzzling private interests can take matters into their own hands and prevent the destruction of their watersheds and the takeover of water delivery systems. Governments, so far, have not taken the lead to protect the water on which their constituents' lives depend. So it will be up to nongovernmental organizations and citizens' groups to change the way water is obtained and distributed and to protect this life-giving resource for coming generations.

THE WAY FORWARD

How Ordinary People Can and
Will Save the Global Water Supply

> *It is done! I am the Alpha and the Omega, the beginning and the end. To the thirsty I will give water without price from the fountain of the water of life.*
>
> — REVELATION 21:6

> *Whose are the forests and the land?*
>   *Ours, they are ours.*
> *Whose the wood, the fuel?*
>   *Ours, they are ours.*
> *Whose the flowers and the grass?*
>   *Ours, they are ours.*
> *Whose the cow, the cattle?*
>   *Ours, they are ours.*
> *Whose are the bamboo groves?*
>   *Ours, they are ours.*
> — NARMADA BACHAO ANDOLAN SONG

I n recent years, an international movement of social advocates, public educators, environmentalists, workers, anti-poverty and human rights activists, debt cancellation proponents, and many others has come together to put human and ecological issues back on the political and economic agenda. Members of this movement are forming powerful alliances with one another to change government policy in their own countries and around the world and to dismantle or reform global financial institutions like the World Bank and international trade agreements such as those generated by the World Trade Organization. They also seek to forge the new international social and environmental contracts that governments need to adopt in order to guard democratic principles and to protect their citizens from the water-depleting dynamics of the new global economy.

As described in Chapter 8, the seeds of the international movement to preserve the world's water have been sown. In many communities around the world, individual fightbacks have stopped the privatization of a local water system, blocked the construction of a dam, or created the impetus to restore a local river or wetland to health. Important as these individual victories are, however, it is now time to build a powerful international coalition of community groups, human rights activists, environmentalists, farmers, Indigenous peoples, public sector workers, and others to save the world's water from theft and pollution and to build a model for a water-secure planet.

The process has already begun. In July 2001, at the beautiful campus of the University of British Columbia overlooking the Pacific Ocean, more than 800 people from 35 countries came together for the first international civil society conference to strengthen the global fight against the commodification of water. Water for People and Nature: A Forum on Conservation and Human Rights was sponsored by the Council of Canadians, a citizens' advocacy organization, and it provided the first opportunity for activists and experts to come together outside the auspices of government, the UN, or the World Bank to talk to one another about shared experiences, beliefs, and plans.

The assembly heard from a wide range of panelists. Public sector workers told of the fight to stop water privatization in their countries.

Scientists shared their expertise and pledged to work with community groups. Environmentalists explained the connection between climate change, clearcutting, and other environmental problems and the world's water crisis. Human rights specialists issued a call for equality in the struggle for water and warned that huge numbers of the world's citizens are already dying from a lack of fresh drinking water. There were powerful personal stories: small farmers fighting foreign-based transnationals in Uruguay that were trying to buy up massive tracts of water-rich land, and South African municipal workers struggling to truly obtain the water rights guaranteed in their constitution.

Two tracks were of particular relevance at the summit. A youth caucus brought several hundred young people from around the world together so they could support one another and take this campaign back into their networks on campuses and in the street. And an Indigenous caucus led by Chief Arthur Manuel of the Interior Alliance of British Columbia brought First Nations peoples from around the world to support one another and to form common strategies in their fight to preserve their ancestral water rights. This caucus endorsed an Indigenous peoples' Declaration of Water, which is now being circulated around the world.

The saddest moment came in a testimony to Colombian Indigenous leader Kimy Pernia Domico, who was supposed to have been present at the conference. On June 2, 2001, Kimy was kidnapped by paramilitary troops thought to be connected to the Colombian government and is now considered "disappeared" and possibly dead. In dedicating the summit to Kimy, the assembly remembered in silence all those who have paid such a terrible price for the land, clean water, and fundamental rights that are taken for granted by some but still denied to millions.

It soon became evident that groups and organizations are already forming in many parts of the world to address the water crisis. Riccardo Petrella, on behalf of leading European intellectuals calling for an international campaign to protect water, shared his powerful dream of a World Water Contract. The Washington-based Public Citizen Water for All project was also warmly embraced. Friends of the Earth International and the International Rivers Network pledged their support for this new movement, and citizens of both Third World countries and the industrialized nations of the North vowed to form alliances and develop common political strategies.

At the end of the summit, the assembly unanimously called for water to remain in "the commons" and launched the Blue Planet Project, a new international civil society movement to protect the world's water. The summit also unanimously endorsed the Treaty Initiative to Share and Protect the Global Water Commons, printed at the beginning of this book. The intention is for the proposed treaty to be signed by the world's governments; in doing so, they would pledge their commitment to protect water as part of the global commons and to administer the earth's fresh water supply as a trust. The Treaty Initiative, it was agreed, will form a key part of the demands of the global citizens' movement in the work leading up to Rio+10 to be held in Johannesburg, South Africa, in the fall of 2002.

While the main focus of the Blue Planet Project is political — reclaiming water as part of the commons and asserting the universal right to water — it is clear that this movement must also address the critical environmental issues around water scarcity. Indeed, water preservation and water equity are the cornerstones of a water-secure world and the starting point for the fledgling citizens' water-security movement.

WATER PRESERVATION

The world's water crisis is grave and should not be minimized. It is going to take massive effort on the part of the majority of nations and citizens to begin implementing the policies and practices that could build a water-secure future. But there *are* solutions. Many community groups, farmers, scientists, and environmentalists are working on proven alternatives.

The single most important tool for a water-secure world is conservation of the world's fresh water supplies and the reclamation of polluted water systems. This will require a change in attitude toward water that will be a challenge to those working for water security. Simply put, humans have to stop thinking that there is an endless supply of water that can be used to attend to our every need and desire. We have to begin changing our ways, in order to meet our water needs with what is available. As Sandra Postel of the Global Water Policy Project says, humanity needs to double water productivity, and soon. That is, we have to get twice as much benefit from each liter of water we remove from rivers, lakes, and underground aquifers if we are to have any hope of providing water for the 8 to 9 billion people who will need it in the next several decades. With technologies known and available today, agriculture could cut its demands by up to 50 percent, industries by up to 90 percent, and cities by one-third, with no sacrifice of economic output or quality of life.

Many North Americans use about 500,000 liters (about 132,000 US gallons) of water each year, at least half of it wasted in lavishly washing cars or leaving taps to drip. Yet people need less than 10,000 liters (about 2,600 US gallons) of water a year to live. In the City of Toronto alone, toilets are flushed 66 million times a day. Californians have 560,000 swimming pools. Common sense reductions in this waste, coupled with infrastructure replacement, could save massive amounts of fresh water all over the world.

In the U.S., all new residential toilets sold since 1994 have to be, by law, high-efficiency, low-flow installations. This has reduced by 70 percent the amount of water needed to flush millions of toilets in American cities. Cities in different parts of the world have saved up to 25 percent of their water through the repair of leaky pipes, the recycling of treated wastewater for urban irrigation, and fines for water waste. Industrial water recycling in western Germany, for example, which began on a large scale in the 1970s in response to anti-pollution legislation, has resulted in so much water conservation that industrial water use has not risen in more than two decades, despite a large increase in the number of factories. American steelmakers, which once consumed 280 tons of water for every ton of steel made, now use only 14 tons of new water. Obviously, these are good-news stories among many other examples of industrial operations polluting indiscriminately. Legislation similar to these laws passed by Germany and the U.S. are needed to control both pollution and water waste in such industries as mining and high-tech.

Environmental analysts Sandra Postel, Peter Gleick, and others have documented the technologies and practices that could be used fairly easily to promote water conservation in agriculture. Heavily subsidized irrigation practices that allow nonsustainable crops to be grown on arid land must be halted. This would promote better farming practices such as growing water-intensive crops only in areas with a lot of natural water in the soil. In addition, the evidence of the harm done to water, animals, and humans by factory farming is mounting and incontrovertible. Legislation to ban or at least strongly regulate factory farms is urgent. Legislation is also needed at all levels to ban or control the use of pesticides, herbicides, antibiotics, chemical nitrates, and fertilizers. In fact, the whole move to corporate farming cannot be water-sustained, and governments everywhere must be called upon to regulate against this trend. At the same time, international policies to promote small-scale farming must be fostered.

Massive leakage of water from inefficient irrigation systems all over the world could be easily and dramatically improved by new and more efficient technologies, as well as by better management and farm practices, — including highly efficient sprinklers and drip irrigation. Drip systems that replace flood systems deliver water directly to individual root plants, eliminating evaporation — thus cutting down on salt build-up and saving water and energy. Drip irrigation is 95 percent efficient, in that almost all the water goes directly to the plant, compared with traditional irrigation, which loses up to 80 percent of its water in evaporation or runoff. Only one percent of the world's irrigated lands now use drip technology, so the conservation potential is great in this area.

For millions of poor small farmers and peasants in the Third World, drip irrigation and other small-scale technologies custom designed for small plots are the only tools that will distribute water in an equitable and sustainable way. As a result, farmer-controlled, small-scale farming is being increasingly looked to as a water-sustainable model for food production. Traditional water collection techniques such as rooftop or mountain-slope water harvesting have also proven to be superior to Western-introduced high-maintenance technology. As the damage from large-scale dams and water diversions becomes more evident, it will become imperative to support small-scale, efficient, and affordable water-harvesting methods around the world. In Nepal, for example, Farmer Managed Irrigation Systems (FMISs) now account for 70 percent of all irrigated crop production in that country. FMISs manage local water resources for the benefit of the whole community and are based on a reverence for Indigenous knowledge and practices, which they incorporate into all phases of the irrigation cycle. Other countries are looking to the Nepalese system for guidance as flood irrigation has decimated many of their farms.

Hand in hand with support for these kinds of agriculture alternatives must be an increased rejection of large-scale dams and diversions. Rivers that once flowed to the sea must be freed again to enrich watersheds, create habitat for aquatic life, and sustain the rich spawning grounds where fresh water meets seawater. While this will take many years, Nature will assist us if we do no more than stop building new dams, whose economic and ecological future is already precarious. Small-farm advocate and conservationist Wendell Berry has put the matter in these poetic terms: "Men may dam it and say that they have made a lake, but it will still be a river. It will keep its nature and bide its time, like a caged animal alert for the slightest opening. In time, it will have its way; the dam, like the ancient cliffs, will be carried away piecemeal in the currents."

Equally, we know that looking at the larger ecological needs of a whole system can give us the answers to water shortage. Watershed planning based on soil water conservation in drought-affected regions of central India has increased crop production and reduced hunger in the communities involved. The UN has designated several areas around the world as Biosphere Reserves — terrestrial and coastal ecosystems where watersheds must be conserved while still being used by humans in a sustainable way. These Biosphere Reserves are characterized by a holistic approach to the protection of Nature and human development through the promotion of cooperation at local, regional, and international levels. As just one example, Friends of the Earth International and others are calling for the whole Dead Sea Basin to be declared a World Heritage Site and Biosphere Reserve in order to save the Dead Sea itself.

In South Africa, where the population is growing four times faster than the water supply, the government has undertaken a major experiment in biosphere planning by linking the reclamation of water with the social and employment needs of the local people. Some remedy must be found because the present situation is desperate. South Africans, for instance, have only half the fresh water per person that they had in 1960, and in 50 years, half the country's rivers will run dry. While a number of water stresses account for this reality, one major source of water depletion in that country was introduced by humans. When early European settlers arrived in South Africa, they missed the trees and parks of their homeland and would scatter tree seeds behind them on their walks. Soon, water-guzzling pine and eucalyptus trees replaced the local plant life that used little water, and rivers started to dry up. Through a 20-year national project called "Working for Water," forty thousand South Africans are now clearing invasive species from their forests and grasslands. Most are from the local poor communities where unemployment was very high. By reclaiming their environment and their fresh water rights and by engaging in digni-fied work, these South Africans are living proof that humans and Nature can co-exist if both are treated with care and respect.

Such respect for Nature must become a core goal of the global move-ment to save water. Fresh water systems cannot survive if the surrounding habitat is destroyed by clearcutting, wetland destruction, and careless urbanization. Integrated environmental policies must become the corner-stone of a series of legal precedents at every level of government, and humanity must specifically reserve water for Nature. In some cases, this will mean reclaiming overstretched river systems from cities and agribusi-ness operations in order to give them back to Nature and the smaller rural communities that are suffering for lack of water.

Finally, one basic law of Nature must be addressed: we cannot continue to mine groundwater supplies at a rate far greater than natural recharge. If we do, there will be, quite simply, no water for our children. The rule of nature is straightforward. Extractions cannot exceed recharge. But many governments in the world have not even begun to ask questions about the location and size of their groundwater reserves, let alone fashioning a policy to preserve them. These calculations need to be made — and they are not sophisticated. As with a bathtub, if water drains away and no or little water is poured in, in time, the tub will be dry as a bone.

WATER EQUITY

The world has recently celebrated the 50th anniversary of the 1948 United Nations Universal Declaration of Human Rights. This declaration marked a turning point in the long international quest to assert the supremacy of human and citizen rights over political or economic tyranny of any kind. Together with the International Covenant on Economic, Social, and Cultural Rights and the International Covenant on Civil and Political Rights, the declaration stands as a 21st-century Magna Carta. Besides granting full human rights to every person on earth regardless of race, religion, sex, and many other criteria, the declaration includes the rights of citizenship — those services and social protections that every citizen has a right to demand of his or her government.

These include social security, health, and the well-being of the family, including the right to work, decent housing, and medical care. The covenants bind governments to accept a moral and legal obligation to protect and promote the human and democratic rights outlined in the declaration, and they contain the measures of implementation required to do so. The individual rights and responsibilities of citizens as established by the declaration, together with the collective rights and responsibilities of nation-states as established in the covenants, represent the foundation stones of democracy in the modern world.

Yet half a century later, the lack of access to clean water means that well over one billion people are being denied one of the basic rights guaranteed in the covenants. Over those 50 years, the rights of private capital have grown exponentially, while those of the world's poor have fallen off the political map. It is no coincidence that the deterioration and depletion of the world's water systems have taken place concurrently with the rise in power of transnational corporations and a global financial system in which communities, Indigenous peoples, and smaller farmers have been disenfranchised. An ecologically sound, water-sustainable future will not happen if the world's water is put on the open market for sale to the highest bidder. If citizens lose control of this precious part of the commons, we will lose the ability to set the conditions according to which fresh water can be preserved and equitably shared.

The emphasis of any citizens' water-security movement must be on the provision of basic water rights for all — a stance that entails vigorous opposition to privatization of the world's fresh water resources. Governments must get the message that they have to assume responsibility to protect water and to provide it to all their citizens as a basic human right. If governments need to price water in order to preserve it from careless waste, the practice must be done within a public system in which the proceeds go not to shareholders or corporate CEOs, but to water reclamation and infrastructure repair, and universally accessible water supply. Riccardo Petrella suggests the following guidelines for a fair water-pricing system: that communities themselves determine local water needs; that any pricing take place only after basic water needs for all are met; that all households and organizations, public and private, be required to make lump sum payments to a community water fund, the amount to be determined on the basis of resources; that the unit price of water per household or organization rise steeply after a certain threshold of sustainability has been exceeded; and that water use beyond the limits the community has set be subject to punitive sanctions. In addition, commercial and industrial users that consume huge amounts of water would be heavily taxed for it and prohibited from passing on such taxes to consumers. However, no one, be it corporation or community, would be allowed to buy water in order to abuse it. A city like Las Vegas, for instance, which wastes enormous amounts of water in swimming pools, colossal fountains, and water imagery would not be allowed to continue depleting local water resources for profit.

Just as the global water-security movement must make water justice a principal goal, so too should it fight to keep water services in public hands. Where the private sector is already involved, however, or in cases where communities choose to work with the private sector, strict guidelines will be needed to protect public health, decent working conditions, and equitable water distribution. Peter Gleick and his colleagues at the Pacific Institute for Studies in Development, Environment, and Security suggest some useful guidelines in these cases.

All agreements with a private company, they say, must guarantee basic free water services to all residents of the community as well as providing for the basic needs of the local ecosystem. Rates must be fair and transparent, and they must act as incentives for users to conserve water (by setting higher prices for profligate water use, for instance). Governments must retain or establish public ownership of water sources and infrastructure and only public agencies should monitor water quality and enforce water-quality laws. Finally, any contract would have to ensure local community participation and oversight.

These are stringent requirements for any private sector water player and would likely act as a disincentive to many of the current water trans-nationals. Instead, they would encourage private enterprises to adhere to ethical water management practices, and these principles could be applied to a number of other sectors.

TEN STEPS TO WATER SECURITY

Armed with this wide range of potential environmental and human solutions to the global fresh water crisis, it is now time to advance an international water-security movement agenda to protect water and to defend it from commercial exploitation. The following are guidelines that should be followed if we are to protect and conserve our scarce water resources and distribute them in a fair and ecologically responsible way:

1. Promote "Water Lifeline Constitutions."
2. Establish local "Water Governance Councils."
3. Fight for "National Water Protection Acts."
4. Oppose the commercial trade in water.
5. Support the anti-dam movement.
6. Confront the International Monetary Fund and the World Bank.
7. Challenge the lords of water.
8. Address global equity.
9. Promote the "Water Commons Treaty Initiative."
10. Support a "Global Water Convention."

1. Promote "Water Lifeline Constitutions."
Water is a global human and ecological asset, and all communities have

the right to use this resource for their needs. The first order of business for civil society is therefore to declare water a public trust to be guarded for all time by our elected leaders. Management of this trust must be the shared responsibility of communities of citizens and of governments at all levels.

Every person in the world should be guaranteed a "water lifeline" of at least 25 liters (about 6.5 US gallons) of free clean water each day as an inalienable political and social right. This must be guaranteed by national and international law and made possible by a combination of three measures: strict water preservation practices, enforced by legislation; taxing those who benefit from excessive water use, such as agribusiness, mining firms, and high-tech water guzzlers; and pricing careless or unduly high water use.

The movement should work toward a Water Lifeline Constitution in every country in the world, in which each national government would guarantee a minimum supply of water to every one of its citizens. Governments would use their powers to ensure the equitable distribution of water to all communities, especially the poor, and to set up educational programs to make people aware of their constitutional rights to water.

2. Establish local "Water Governance Councils."

The best advocates for water are local communities and citizens, as water-endangering practices are most easily observed and felt at the local level. So it is crucial for these parties to become stewards of their local water systems as equal partners with governments. Local communities must establish co-governing structures with elected citizens and local government water authorities to jointly oversee wise water management practices. Local Water Governance Councils, for instance, could monitor and protect local water supplies, observe local farming practices, and report on polluting industries. They could oversee local watershed governance systems and implement practices emphasizing "local people and farmers first," whereby local communities have first rights to local water. They could sponsor independent, publicly funded research into all aspects of policy governing water. And they could work with local Indigenous groups to support their self-governance rights to waters in their territories.

Most important, local citizens' groups could oppose the privatization of local water services and advocate in favor of what Public Services International (PSI) calls "PUPS," or public-public partnerships — twinning arrangements which link established public sector providers, such as municipalities, government aid agencies, and public sector unions, with public water services in need of financing or restructuring. PUPS are the alternative to PPPs, or public-private partnerships, in which water service restructuring is done by allowing a private company to run the local water system. Local Water Governance Councils could ensure that such transactions are transparent and accountable and that the PUPs are delivering water in an equitable fashion. These councils could also oversee any contracts between local governments and the private sector to ensure that public control is maintained at all times.

3. Fight for "National Water Protection Acts."

The water-security movement has a right to demand strict nation-state laws and regulations to protect fresh water resources and guarantee water for every citizen. Each nation's National Water Protection Act should address the following issues:

- A *Water Lifeline Constitution* guaranteeing every citizen access to clean water and sanitation services (see Step 1);
- *Water Pricing*, including establishing strict conditions on the pricing of water based on equity, universality, higher fees for agribusiness and industry, and sanctions against water abuse;
- Support of *Public Water Services*, including legislation against handing over control of municipal water services to the private sector and setting stringent conditions on any private sector involvement in the delivery of water or wastewater services;
- *Water Preservation*, including setting aside large tracts of aquatic systems for protection. This includes the need to support watershed management and to create regulatory frameworks to protect watersheds;
- *Water Conservation*, including targets for industry, agriculture, and cities, including a full infrastructure repair program with tight timeframe commitments. Limits on groundwater extractions should also be set;
- *Water Reclamation*, including the cleanup of polluted river systems and

wetlands. Related issues such as clearcutting, global warming practices, and factory farming, would require strict government action;

- Consideration of other *Water Stakeholders*, such as aquatic species and future generations. No decisions about water use should ever be made without a full consideration of the impacts to nonhuman species, the ecosystem, and the future needs of humans;

- *Drinking Water Testing and Standards*, including federal legislation to establish national safe water standards for all communities in the country;

- *Water Standards for Industry and Agribusiness*. Every level of government must commit itself to creating and enforcing strict laws against industrial dumping, the use of pesticides, and the discharge of toxins into waterways or landfills;

- *Water-Friendly Technology*, including alternative sources of power like solar energy and alternatives to mega-power projects, like dams, diversions, and hydroelectric facilities; and

- *Restrictions on Bottled Water*, including limiting water extractions, setting environmental standards, charging companies high rate structures for water extraction, and giving preferential access to local bottling companies that will guarantee local job creation.

4. Oppose the commercial trade in water.

The citizens' water-security movement must take a clear and unequivocal stand against the commercial trade in water. This industry is still in its infancy and can be stopped by massive public opposition and strong government legislation. Governments must enact laws that ban the commercial export of bulk water by tanker, water bags, and diversion. Such laws would not apply to traditional small-scale water-trading arrangements between farmers and communities. They would apply to large-scale commercial trade in water, run by corporate interests.

Governments must also set stringent conditions on the burgeoning bottled water industry. The job of government is to ensure safe, publicly supplied drinking water so that citizens are not forced to pay high prices for boutique water in bottles. The practice of selling local water supplies to large corporate bottlers also disenfranchises farmers and local communities and must be strongly regulated or banned.

Civil society groups should also address the issue of water sharing in times of crisis and encourage public debate and dialogue concerning this crucial question. They must demand independent research on the effect on ecosystems of bulk water diversion to determine how much water can be transferred out of a watershed in emergency situations while still ensuring the continued health of the ecosystem.

Water as a good, a service, and an investment must be exempted from all existing free trade agreements, including the World Trade Organization, the North American Free Trade Agreement, and all Bilateral Investment Treaties (BITs) between nations. In addition, water must not be included in upcoming negotiations on services, such as those surrounding the General Agreement on Trade in Services (GATS) and the Free Trade Area of the Americas (FTAA). Civil society water proponents must work with other groups in challenging the current mandate of these agreements and in opposing them altogether if their mandates are not substantially changed.

5. Support the anti-dam movement.

Any civil society movement to protect the earth's fresh water sources needs to support the work of groups like International Rivers Network and others who are fighting to bring the dam industry under democratic control. This means teaming up on common projects, strategies and campaigns. It also means endorsing the 1988 *San Francisco Declaration of the International Rivers Network: The Position of Citizens' Organizations on Large Dams and Water Resource Management*. This declaration sets conditions for the construction of any dam, including: transparency of process; exploration of more environmentally sound alternatives; environmental, social, and economic impact assessments; accountability to the local people who have the right of veto; full financial compensation to displaced persons; ecosystem protection; protection of local food supplies; guarantee of local health protection; and the inclusion of environmental and social costs in any economic forecasts. These basic themes were echoed in the November 2000 *Report of the World Commission on Dams*, which should be endorsed as well.

In 1994, 326 groups and coalitions in 44 countries endorsed the *Manibeli Declaration Calling for a Moratorium on World Bank Funding of Large*

Dams, named for the heroic resistance of the people of the village of Manibeli in India's Narmada Valley. This declaration called for an immediate moratorium on all World Bank funding of large dams until the Bank carries out the following measures: establish a fund to provide reparations to displaced people; guarantee no forced resettlement in countries without the capacity to ensure full livelihood and human rights protection; agree to evaluate all existing large dams and their environmental and social costs; integrate all World Bank projects into locally approved comprehensive river basin management plans; and allow for independent monitoring and auditing of all projects.

6. Confront the International Monetary Fund and the World Bank.

Similar preventive and restrictive measures must be established with regard to International Monetary Fund and World Bank-sponsored water privatization projects. At the moment, these institutions are promoting water privatization schemes in many countries with the full support of funding governments, aid agencies, and the United Nations. Therefore, the civil society water protection movement must call for a moratorium on all IMF/World Bank water privatization schemes until a set of strict conditions can be put in place to ensure total public control over all water service projects in every country where these institutions operate.

A group of American NGOs, including Fifty Years Is Enough, Results, Globalization Challenge Initiative, Center for Economic Justice, Center for Economic and Policy Research, the Quixote Center, and Essential Action, have drafted proposed U.S. legislation that would cut off U.S. federal money for World Bank/IMF water privatization that does not include a water lifeline for every citizen. The proposed draft legislation also calls for the termination of projects that eliminate subsidies for water consumers and public subsidies for the water system overall. In addition, an amendment to the legislation authorizing Congress to fund the IMF and the World Bank has been proposed by Representative Jan Schakowsky. If this amendment was passed, it would stop the U.S. from continuing to support policies at the IMF and the World Bank which have resulted in poor people losing access to clean drinking water. Similar legislation in other donor countries could be drafted.

Eventually, strict conditions will have to be imposed on the IMF and the World Bank to ensure that they oversee any water projects in which they are involved that also involve the private sector. As stated earlier, such agreements will have to guarantee basic free water services to all residents of the community as well as providing for the basic needs of the local ecosystem. Rates will have to be fair and transparent. Governments will have to retain or establish public ownership of water sources, and only public agencies will be able to monitor water quality and enforce water-quality laws. Any contract will also have to ensure local community participation and oversight.

7. Challenge the Lords of Water.

It is clear that the work of this movement cannot succeed as long as transnational water corporations continue to work in ways that have enormous influence on government and international policy. At present, the water lords have little difficulty gaining the attention and approval of governments as they promote the privatization and commodification of the world's fresh water supplies. To counteract this powerful lobby, civil society must address a whole range of mechanisms to ensure that these private entities do not damage, deplete, or destroy local water sources and waterways by subjecting them to their quest for ever-increasing profits.

One of the first steps is research; much more needs to be known about these corporations and their influence on governments — local and national — the media, the United Nations, the WTO, the IMF, and the World Bank. What is the relationship between water delivery corporations and the public funds funneled through these global financial institutions? How do major water bottling corporations become so influential in the halls of political power that they affect nation-state legislation and trade rules to their benefit?

Citizens must also demand local and national laws that will restrain and regulate the activities of these water giants within their own boundaries. In keeping with several UN covenants, nation-state governments have not only the ability, but also the responsibility to regulate foreign investment and transnational corporations in order to compel them to serve the social and environmental needs of local communities. Legislation must also put an end to water-access preference and subsidized water rates for corporations.

Clearly, another goal must be to break the financial influence that corporations can have on governments through corporate donations to politicians. Corporate access to the inner sanctums of institutions like the WTO and the World Bank must also cease. The United Nations must stop giving preferential treatment to companies like Suez and Vivendi and once again take up the cause of ordinary people. And governments must start taxing corporations again in order to pay for the water preservation and water equity projects outlined here. Governments must unite, perhaps under the banner of the UN, to shut down tax havens for corporations.

Finally, the struggle to bring democratic control to transnational water companies must be part of a larger quest to bring the rule of law to global capital in general. Citizens working on the water issue will want to work with groups calling for the "rechartering" of corporations — the campaign to re-establish the once-held notion that society grants corporations the right to operate and that society can revoke their charters if they are not behaving as good corporate citizens.

8. Address Global Equity.
There will be no water equity until the social inequities of our world are addressed. Economic globalization, for instance, with its passionate adherence to the goal of unlimited growth is neither just nor water sustainable. Water advocates must therefore work with groups and organizations dedicated to closing the world's poverty gap and reclaiming the commons in a whole host of areas. This means cooperating with the anti-globalization movement and its work to create new international instruments to promote fair trade rules, equity-based investment systems, and environmental and human rights agreements with enforcement mechanisms.

It is also crucial for the water-security movement to bring environmental concerns and partners together with social justice concerns and partners. Otherwise, it is entirely possible that the two will start to conflict with each other in a political environment of scarcity. Already some environmentalists, understandably desperate to curtail the waste of the world's fresh water, are calling for across-the-board cost-recovery pricing systems without taking into account the effect that this would have on the poor. This could set the stage for a deep split between these environmentalists and anti-poverty and human rights groups.

Finally, the water-security movement must advocate hard solutions to the terrible economic imbalance between the industrialized and non-industrialized nations. Working with groups like Fifty Years Is Enough, it must call for the end of the Structural Adjustment Programs (SAPs) imposed by the World Bank and the IMF that have forced so many Third World countries to abandon universal health and education services. Another organization that should be involved is Jubilee 2000, the international ecumenical coalition founded on the biblical law of restoring lands and forgiving debts every 50 years. Jubilee 2000 and others are calling for the cancellation of the onerous Third World debt that prevents many countries from putting resources into basic sanitation and water-service delivery. If these debts were canceled, their economies would also have a chance to prosper, so they would no longer need to be objects of charity from the nations of the North.

Foreign aid budgets of the industrialized countries must be restored to their previous standards (0.7 percent of GDP). Many of the same powerful countries of the North who are backing a commercial future for water resources have decimated their foreign aid budgets in recent years. And the movement must join with the Halifax Initiative and others calling for a Tobin Tax — a global tax on financial speculation that could help pay for water infrastructure and universal water services.

9. Promote the "Water Commons Treaty Initiative."

In September of 2002, thousands of people will converge on leafy Sandton, the wealthy suburb of Johannesburg, South Africa, for the World Summit on Sustainable Development (Rio+10). The irony of the site is not lost on South African water activists; Sandton is the Third World's richest large suburb, and the South African government is building a gargantuan luxurious complex for the expected guests that will include shopping malls, exclusive restaurants, and movie theaters. But Sandton, and its huge estates with English gardens and swimming pools, is right next door to Alexandra Township, one of the poorest communities on the African continent. Between the two runs a river so polluted that it has cholera warning signs along its banks.

This setting will serve as a poignant backdrop for the politics expected at Rio+10. Many observers are expecting a powerful push by the World Bank, the WTO, cash-strapped governments, and the United Nations, prompted by corporate interests, to turn to the private sector for answers to the world's environmental problems. Citing massive failure to accomplish the goals set out at the first Rio Summit, many key players are likely to declare that governments are just not up to the task of cleaning up the world's pollution and will call upon the private sector to step up to the plate. On no issue is this more likely to happen than on water, already declared by South African Environment Minister Valli Moosa to be one of the two central themes of the summit, along with climate change.

The global citizens' water-security movement must be ready to counteract this move to have Rio+10 give its approval to a process of commodification of the world's fresh water supplies. The Treaty Initiative to Share and Protect the Global Water Commons, printed at the beginning of this book and launched by the Blue Planet Project at the Vancouver Summit in the summer of 2001, is designed to be the primary focus of a fightback on this front at Rio+10. It states in clear terms that water is a public good and a human right and must not be appropriated for profit. It asks that, in signing, governments and Indigenous peoples agree to administer the world's water as a trust. This treaty is integral to the new water movement; it is an excellent starting point from which national groups and coalitions can launch campaigns. In every country and starting immediately, governments must be convinced to agree to adopt the treaty when it is presented by the citizens' water-security movement at Rio+10 in Johannesburg. At the same time, NGOs must begin to work together to give full support to the water-security movement.

10. Support a "Global Water Convention."

Finally, as the Global Water Contract so clearly points out, the water-security movement must call for the creation of international instruments and institutions to protect the world's shared fresh water supplies. As it stands now, says Riccardo Petrella, there is a deplorable lack of world bodies with sufficient power to provide a clear sense of direction and to monitor the implementation of existing conventions. Neither is there a world body of law that can begin to deal with the myriad of environmental, social, and jurisdictional problems related to water.

Therefore, there is an immediate need for new agreements to promote water as a global commons and a public trust and for a new body of international law to be formed, based on the twin cornerstones of water preservation and water equity. A legally binding Global Water Convention would

- adopt the Treaty Initiative to Share and Protect the Global Water Commons;
- integrate the right to water into the United Nations' Universal Declaration of Human Rights and other existing charters and conventions dealing with the rights of women, children, and Indigenous peoples;
- establish adequate forms of water management on a global scale;
- draft a body of international law that would enforce the principles of the Treaty Initiative to Share and Protect the Global Water Commons;
- set global targets for access to water for all persons;
- coordinate nation-state laws to protect, preserve, and reclaim water;
- create new legally binding environmental treaties to protect the world's water from pollution and exploitation; and
- bring together parliamentarians from all countries to settle water disputes through the principles of equitable distribution and "peace through water equity."

It is clear that a Global Water Convention would lead to the creation of a permanent new international body to oversee this ambitious program. How this might happen and how, in fact, the creation of such a convention in the first place could take place, depends on how flexible the Rio+10 Summit turns out to be. It would be the ideal venue for the launch of a process leading to the creation of the convention. Other targets for both the Treaty Initiative and the Global Water Convention are the third World

Water Forum to be held in Kyoto, Japan, in March of 2003 and the fourth World Water Forum, to be held in Montreal, Canada, in March 2006.

Readying the global citizens' movement to challenge such a powerful body, with a totally different mandate, is a formidable prospect indeed. Yet it is absolutely necessary, and the Montreal forum gives the movement a five-year target to turn the dominant global water agenda around. Planning must start immediately to determine how to bring large numbers of like-minded groups to these meetings, how to put the movement's issues and perspectives on the agenda, and how to enlist the support of the vast majority of the delegates to these meetings who share the view that water is part of the global commons and who would join in a citizens' campaign if given the opportunity.

It must be a key goal that by the fourth World Water Forum, the tide of opinion will have changed and governments and the United Nations will be working with citizens' organizations to announce the adoption of the Water Commons Treaty and the creation of the Global Water Convention.

Eleanor Roosevelt once said, "The future belongs to those who believe in the beauty of their dreams." The growing number of citizens and groups around the world who belong to the Blue Planet Project and other organizations fighting for a water-secure future believe in the beauty of this dream: that our global water crisis will become the source of global peace; that finally humanity will bow before Nature and learn to live at peace within the limits Nature gives us and with one another; and that through our work together, the peoples of the world will declare that the sacred waters of life are the common property of the earth and all species, to be preserved for all generations to come.

NOTES

CHAPTER 1: RED ALERT

A number of influential institutions, including the United Nations, Worldwatch Institute, the Pacific Institute for Studies in Development, Environment and Security, and the World Bank have documented world water scarcity in detail. We are particularly indebted to Peter Gleick's *The World's Water: The Biennial Report on Fresh Water Resources, 1998-1999*, published by Island Press in Washington. The World Resources Institute's *Comprehensive Assessment of the Freshwater Resources of the World*, edited by the distinguished professor I.A. Shiklomanov and published by a number of United Nations institutions, including UNDP, UNEP, UNESCO, and the World Health Organization (WHO), is an excellent source of data. So is the United Nations Commission on Sustainable Development's 1997 report, *Comprehensive Assessment of the Freshwater Resources of the World*. Allerd Stikker's Ecological Management Foundation in the Netherlands also produced a helpful paper called *Water Today and Tomorrow*, for the 1998 Futures series published by Pergamon Press.

Time magazine's November 1997 special on the environment contains excellent material, as does the *National Geographic's* 1993 special edition on North America's fresh water situation, called *Water: The Power, Promise, and Turmoil of*

North America's Fresh Water. The New York Times published a very good special feature on water in its December 8, 1998, edition, written by William K. Stevens. Lester Brown of the Worldwatch Institute describes the China crisis in his 1995 book, *Who Will Feed China? Wake-Up Call for a Small Planet*, published by W.W. Norton, as well as in the July 1998 issue of *World Watch* magazine.

E.C. Pielou, a Canadian scientist and naturalist, provided immense knowledge about the everyday workings of water in her 1998 book, *Fresh Water*, published by the University of Chicago Press. We are also indebted to Michal Kravcík and his team of scientists at the NGO People and Water for their groundbreaking work, first published in Slovakia as an in-depth study and then, in 2001, compiled in a smaller publication for English audiences entitled *New Theory of Global Warming*. In his 1966 book, *The Lion*, Robert S. Strothers predicted the Mexico City water crisis, and Linda Diebel of the *Toronto Star* chronicled that terrible story in a May 1999 series. Finally, we appreciated Marq de Villiers' award-winning book *Water* published in 1999 by Stoddart Publishing of Toronto.

CHAPTER 2: ENDANGERED PLANET

Everyone working on the issue of water owes a huge debt of gratitude to Sandra Postel of the Global Water Policy Project in Amherst, Massachusetts. Her writing is abundant, but we would like to especially credit three publications: her 1992 book, *Last Oasis: Facing Water Scarcity*, published by W.W. Norton; her 1996 paper for the Worldwatch Institute, *Dividing the Waters: Food Security, Ecosystem Health, and the New Politics of Scarcity*; and her 1999 book, *Pillar of Sand: Can the Irrigation Miracle Last?* , also published by Norton. We are also grateful to Janet Abramovitz for her work, particularly *Sustaining Freshwater Ecosystems*, published in the Worldwatch Institute's 1996 *State of the World* series.

The Nature Conservancy has published several helpful papers available from their Arlington, Virginia, offices. Particularly useful were the 1998 *Rivers of Life: Critical Watersheds for Protecting Freshwater Biodiversity* and the 1996 *Troubled Waters: Protecting Our Aquatic Heritage*. Elizabeth May of Sierra Club of Canada has written a comprehensive account of the destruction of Canada's forests and the dangers of clearcutting in her 1998 book, *At the Cutting Edge: The Crisis in Canada's Forests*. We also relied on the August 2001 United Nations report *An Assessment of the Status of the World's Remaining Forests*. Simon Retallack and Peter Bunyard co-edited an excellent special edition of *The Ecologist* in March 1999 on climate change. The Institute of Political Ecology in Chile provided information in its 1996 publication, *The Tiger without a Jungle: Environmental*

Consequences of the Economic Transformation of Chile.

We are grateful to several publications and organizations for detail on the crisis of the Great Lakes. In 1997, the Canadian Environmental Law Association and Great Lakes United published *The Fate of the Great Lakes: Sustaining or Draining the Sweetwater Seas?* The International Joint Commission has also published many studies. We particularly note their February 2000 report, *Protection of the Waters of the Great Lakes: Final Report to the Governments of Canada and the United States.* Environmental writer and researcher Jamie Linton prepared an excellent study for the Canadian Wildife Federation called *Beneath the Surface: The State of Water in Canada.*

Sally Deneen collected a great deal of information on wetlands published in a December 1998 article called "Paradise Lost: America's Disappearing Wetlands" for *E Magazine.* On the issue of factory farms, we want to thank Mark Ritchie of the Institute for Agriculture and Trade Policy in Minneapolis and David Brubaker with the Center for a Livable Future, Johns Hopkins University, Maryland, for their tireless work and impeccable research.

CHAPTER 3: DYING OF THIRST

Anne Platt of the Worldwatch Institute provided a rich source of information on how water carries disease. Most particularly, we note her March 1996 publication for Worldwatch called *Water-Borne Killers.* We are grateful to the U.S. branch of Physicians for Social Responsibility, which published *Drinking Water and Disease: What Health Care Providers Should Know*, in 2000, and to the Sierra Legal Defence Fund in Canada for its 2001 report, *Waterproof: Canada's Drinking Water Report Card.*

Klaus Topfer of the United Nations Environment Programme sounded the alarm on water shortages in a speech at a March 1998 conference, the International Conference on Water and Sustainable Development. We also relied on the latest annual reports of the United Nations Development Programme, the *United Nations Human Development Report*, which measures human well-being in a number of areas, including access to water and sanitation. On the water crisis in South Africa, we used a number of sources, including *Drought and Liquidity: Water Shortages and Surpluses in Post-Apartheid South Africa*, Patrick Bond and Greg Ruiter's 2000 review published for the Pretoria-based Human Sciences Research Council.

Also helpful was Jacques Leslie's special feature on water "Running Dry: What Happens When the World No Longer Has Enough Fresh Water?" for the July 2000 edition of *Harper's Magazine.* It was Leslie who noted the 1996 comment from Colin Powell, chairman of the U.S. Joint Chiefs of Staff, indicating

Notes

that during the 1991 Gulf War, the United States considered bombing dams in the Euphrates and Tigris rivers north of Baghdad but backed off for fear of high casualties. Another excellent *Harper's* piece was the June 1998 article by Wade Graham, "A Hundred Rivers Run through It: California Floats Its Future on a Market for Water." On the high-tech sector in California and elsewhere, we are indebted to the Campaign for Responsible Technology in San Jose and the Silicon Valley Toxics Coalition. The Interior Alliance of British Columbia and the Council of Canadians produced an excellent 2001 report on water and First Nations called *Nothing Sacred: The Growing Threat to Water and Indigenous Peoples*.

There is also a great deal published on the impact of large dams. We were grateful for the work of the World Commission on Dams and for its final 2000 report, *Dams and Development: A New Framework for Decision Making*. In addition, there is excellent material from a group of leading scientists in the November 1999 report of the World Conservation Union, *Large Water Impacts on Freshwater Biodiversity*. Most particularly, however, we want to acknowledge the groundbreaking work of the International Rivers Network and its regular publication, *World Rivers Review*, as well as the tireless work of its campaign director, Patrick McCully. His 1996 book, *Silenced Rivers: The Ecology and Politics of Large Dams*, is a masterpiece.

CHAPTER 4: EVERYTHING FOR SALE

The basic framework for this chapter was shaped by several sources. Jerry Mander's introduction to *The Case against the Global Economy* (London: Earthscan Publications, 2001) identifies the basic elements and forces of economic globalization (pp. 1-18). For a description of the Washington Consensus, see Maude Barlow and Tony Clarke's *Global Showdown* (Toronto: Stoddart, 2000), pp. 57-58ff. Analysis of the Trilateral Commission is found in Patricia Marchak's *The Integrated Circus: The New Right and the Restructuring of Global Markets* (Montreal and Kingston: McGill and Queen's University Press, 1993); David Korten's *When Corporations Rule the World* (San Francisco: Kumarian Press Inc. and Berrett-Koehler Publishers, Inc., 1995); and Tony Clarke's *Silent Coup: Confronting the Big Business Takeover of Canada* (Ottawa and Toronto: Canadian Centre for Policy Alternatives and James Lorimer, 1997). The statistics on the top 200 transnational corporations in the world provided by the Institute for Policy Studies are found in Sarah Anderson, John Cavanagh, and Thea Lee's *Field Guide to the Global Economy* (New York: The New Press, 2000). And the quote from John McMurtry was taken from his article "The Meta Program for Global Corporate Rule" published in *The CCPA Monitor* (June 2001) by the Canadian Centre for Policy Alternatives.

254

In the section "Commodifying Nature," reference is made to Herman E. Daly and John B. Cobb's classic work, *For the Common Good: Redirecting the Economy toward Community, the Environment and a Sustainable Future* (Boston: Beacon Hill Press, 1989). For a discussion of Daly and Cobb's thesis in relation to globalization issues, see William Greider's *One World Ready or Not: The Manic Logic of Global Capitalism* (New York: Simon & Schuster, 1997), pp. 451-459. Vandana Shiva's insights are taken from her *Stolen Harvest: The Highjacking of the Global Food Supply* (Cambridge, MA: South End Press, 2000). The examples from India on the commodification of water and the notion of "common resource property" are taken from *Licence to Kill?* , published in March 2000 by the Research Foundation for Science, Technology and Ecology in New Delhi. The data on privatization of public services is drawn mainly from David Hall's research at the Public Service International Research Unit based at the University of Greenwich in the U.K. In particular, David Hall's booklet, *Water in Public Hands*, published by Public Service International in July 2001, contains the data used to illustrate the model of "public-private partnerships"(PPPs). The quotation about the kind of financing provided by the World Bank is taken from "Tapping the Private Sector Approaches to Managing Risk in Water and Sanitation," an *RMC Discussion Paper Series 122* of the World Bank by David Haarmeyer and Ashoka Mody, 1998, p. 13.

The data on financial speculation in reference to water was originally outlined in Maude Barlow's booklet *Blue Gold*, published by the International Forum on Globalization in 2000. Analysis of the Cadiz Project, including quotations by Brackpool and Cohelo, are contained in a series of documents provided by the Campaign to Stop the Mojave Water Grab, near Los Angeles. Arguments against the economic and environmental feasibility of the Cadiz Project are outlined in *Desert Report* (published by the Sierra Club in February 2001). The story of George Soros' gamble with John Major was cited in Richard Barnet and John Cavanagh's article "Electronic Money and the Casino Economy," published in *The Case against the Global Economy*. The statistics cited on foreign direct investment were taken from the *World Investment Report 1996*, published by the United Nations, and the figures on export growth and the increasing volume of world trade were taken from several sources, including *World Economic Outlook* (October 1997), *Financial Statistics Yearbook* (1997), and *International Financial Statistics* (1998). Simon Retallack's "The Environmental Cost of Economic Globalization," published in *The Case against the Global Economy*, was very helpful in describing the ecological impacts of international competitiveness, including the data on cash crop exports from Third World countries. The data on tax breaks in Texas and subsidies in New Mexico for the water-guzzling computer industry were taken from the original *Blue Gold* book-let published by the International Forum on Globalization. The analysis of the "corporate security state," as well as the references to Ursula Franklin were out-lined in *Silent Coup*.

CHAPTER 5: GLOBAL WATER LORDS

This chapter is based largely on a study initially prepared by Gil Yaron for the Polaris Institute in Ottawa entitled *The Final Frontier: A Working Paper on the Big 10 Global Water Corporations and the Privatization and Corporatization of the World's Last Public Resource* (March 2000). The research on the major water cor-porations has since been updated at the Polaris Institute by Darren Puscas. The introduction, as well as each chapter of the study is available at www.polarisin-stitute.org. The story of the privatization of water services in Buenos Aires by the Suez-led consortium is documented in two papers prepared by Dr. David MacDonald and Alex Loftus: *Lessons from Argentina: The Buenos Aires Water Concession*, published by the Municipal Services Project, based at Queen's University, Canada, and "Of Liquid Dreams: A Political Ecology of Water Privatization in Buenos Aires," published in *Environment and Urbanization*, vol. 13, no. 2, October 2001. Special thanks goes to them for their work. See also Shawn Tully's "Water, Water Everywhere" in *Fortune*, May 15, 2000. Our analy-sis of world water industry trends (the "Blue Bonanza") was informed by "The Rising Tide of Water Markets" in *Global Water Intelligence* (August 2000).

The portraits of the two global water giants, entitled "Suez's Conquest" and "Vivendi's Empire," are based on corporate profiles prepared by Darren Puscas at the Polaris Institute. Copies of these two corporate profiles are available on the Internet at . Jean-Philippe Joseph's unpublished paper, "Vivendi: anatomie de la pieuvre" (January 2001), is available on request at . Enron as a multisector service provider (including its water company Azurix) is documented in a cor-porate profile prepared by Darren Puscas, which is available at See also "Enron Ponders Its Next Move" in *Global Water Intelligence* (September 2000), for indus-try insights into the story of what happened with Azurix. The analysis of RWE is found in *The Final Frontier*, and the corporate profile of RWE is available at Information on RWE's recent acquisition, Thames Water, including the com-pany's record regarding leakages and related issues is based on news reports available on the Internet and Lexis-Nexis. Similarly, the information on SAUR's water operations is based on a corporate profile of Bouygues available at , as well as more recent news reports available on the Internet.

The information about Suez in La Paz, Bolivia, is from Kristen Komibes' "Designing Pro-Poor Water and Sewer Concessions: Early Lessons from Bolivia

Private Participation in Infrastructure Group at the World Bank, 1999, pp. 30-34. The report on Suez in the United Kingdom appeared in "South West, North West Water Score Lowest for Quality in England, Wales," AFX News, July 7, 1999, available through Lexis-Nexis. Suez in Potsdam appeared in news reports compiled by PSIRU in 1998. See www.PSIRU.OR9/news/4193.htm. Aguas Argentina's workforce layoffs were recorded in Daniel Cieza, "Argentine Labour: A Movement in Crisis," (*NACLA Report on the Americas*, vol. 31, no. 6 (1998), p. 23, and in MacDonald and Loftus's "Of Liquid Dreams," pp. 195-196. Reports on privatization in Jakarta appeared in the *Jakarta Post* (September 18, 1998, and May 13, 1999).

For the most part, the section entitled "Privatized Fiasco" is documented in Gil Yaron's *The Final Frontier* and by the Public Services International Research Unit in several of its publications. For example, the case of Suez in Grenoble is detailed in a PSIRU publication by David Hall and Emanuele Lobina entitled *Private to Public: International Lessons of Water Remunicipalization in Grenoble, France*. The cases involving Vivendi in Angoulême and St. Denis, France, are noted by PSIRU in *Water in Public Hands* (July 2001) and documented by David Hall in "Privatization, Multinationals and Corruption," published in *Development and Practice* (vol. 9, no. 5, pp. 539-556).

Vivendi's operations in Puerto Rico were reported by Interpress (August 16, 1999, and September 16, 1999) and in David Hall, *Water in Public Hands*, PSIRU report, July 2001, p. 10, as well as in "Water Company Nears Collapse" by Carmelo Ruiz-Marrero (May 26, 2001) and "Puerto Rico: Water Company near Collapse," *Black World Today* (May 28, 2001). The Sereuca Space-Vivendi-Tandiran proposed joint venture in Nairobi is described in Peter Munaita's "French Water Deal to Cost Kenyans $25 M" in *The East African*, August 7, 2000, and in Peter Munaita's "Government Halts Vivendi, NCC Water Project," *The East African*, August 20, 2001. Azurix's operations in Bahia Bianca are detailed in "Argentine City Says Tap Water Is Toxic" in *U.S. Water News Online*, May 2000, and in "Azurix Water Bugs Argentina" in *Houston Business Journal*, May 5, 2000. More details on Azurix in Argentina may be found in "BA Governor to Request End to Waterworks Contract — Argentina," *Financial Times Information*, January 15, 2000; "BA Government Softens Line on Azurix-Argentina," *Business News America*, January 19, 2001; "Argentina/Companies — Another Blow for Azurix," *Financial Times Energy Newsletters — Global Water Report*, Feburary 23, 2001; and "Enron's Azurix to Rescind Buenos Aires Province Water/Sewage Contact," *AFX European Focus*, September 7, 2001.

Reports on Bechtel, Enron, and other companies operating in the United Kingdom appeared in "Worst U.K. Polluters Include Enron, Vivendi, Suez-

Lyonnaise" at www.psiru.org/news/3437.htm; "U.K. Environment Agency, 'Names and Shames' Worst Corporate Polluters" in the U.K. Environment Agency's 1999 Hall of Shame (June 1999). Available at http://www.ukenvironment.com/archcorprespon.htm; and the U.S. Environmental Protection Agency's ERNS database http://d1.RTKNET.org/ERN/fac.php. Last updated 12/17/97. (Enter ENRON or BECHTEL under "discharger.")

The World Bank report about the privatization process in general appears in Susan Rose-Ackerman, "The Political Economy of Corruption: Causes and Consequences," World Bank Public Policy for the Private Sector, Note no. 74 (Washington, D.C.: The World Bank, 1996). The description of the efficiency of Chile's public water companies appears in *Private Sector Participation in the Water Supply and Wastewater Sector: Lessons from Six Developing Countries* (Washington, D.C.: World Bank, 1996). The description of the public water utility in São Paulo is found in David Hall's *Water in Public Hands*, p. 17.

CHAPTER 6: EMERGENT WATER CARTEL
This chapter also makes use of quotations and data published in a variety of sources. References by Terrance Corcoran and others to the prospects of a global water cartel were found in the *National Post* in February 1999. The quotation from Robert Kaplan's article "Desert Politics" is taken from the July 1998 edition of *The Atlantic Monthly*. References to water pipeline construction in Turkey, Scotland, and Australia were initially drawn from various news reports and later summarized in Maude Barlow's booklet *Blue Gold*, pp. 22, 23, published by the International Forum on Globalization. For details on the Libyan pipeline project that Gadhafi had constructed through a South Korean conglomerate, see Marq de Villiers' *Water* (Toronto: Stoddart, 1999), pp. 179-182. Richard Bocking's analysis of the use of supertankers to transport water across the Pacific was outlined in his brief entitled, "Water Export and the Multilateral Agreement on Investment," submitted to the British Columbia Special Legislative Committee on the MAI, in October 1998. (Bocking is a Canadian filmmaker and writer who specializes in environmental issues, among other things. He has also written a book *Mighty River: A Portrait of the Fraser* (Vancouver: Douglas & McIntyre, 1997).) The *Alaska Business Monthly* is a key source of information on water export plans from Sitka, Alaska, and related operations of corporations like Global H2O and World Water. Quotes from Fred Paley, as well as references to the limits imposed by the *Jones Act*, are found in an *Alaska Business Monthly* article, "Exporting Alaska's Water" (November 1998).

The political background to the GRAND canal project was outlined in Robert Chodos et al.'s *Selling Out* (Toronto: James Lorimer, 1988), Chapter 2, "Resources:

Redesigning God's Plan." An earlier analysis of these canal schemes from Canada to the U.S. is presented by Richard Bocking in a book called *Canada's Water: For Sale?* (Toronto: James Lewis and Samuel, 1972). Background information on NAWAPA is found in *Selling Out* and in Marq de Villier's *Water*, pp. 327, 330. The story of the emerging water-bag technology is outlined in an article called "Oceanic Answer" in *Water 21: A Magazine of the International Water Association* (February 2000). Further data on companies like Aquarius and Nordic Water Supply were found in various business news reports (January through May 2001) obtained through Lexis-Nexis. The portrayal of the water-bag operations of Medusa and Spragg are based on the *Water 21* article.

On the topic of bottled water, one of the most thorough studies to date is the report of the U.S. National Research Defense Council, *Bottled Water: Pure Drink or Pure Hype?* published in February 1999. Quotations in this section, including the quotation from the CEO of Perrier, are taken from this report. The figures on the annual volume of water being bottled, along with other insights, are taken from the study *Bottled Water: Understanding a Social Phenomenon* (commissioned by the World Wildlife Federation and released in May 2001). Martin Mittelstaedt's articles in the *Globe and Mail*, "Canada's Giving Away Its Precious Water" (September 21, 1999) and "Bottled Water Gushing South but Canada Gets Little in Return" (September 22, 1999) provided useful background information on the bottled water industry's access to water supplies in Canada. An update on the bottled water industry is included in an article entitled "Multinationals Tap into the Bottled Water Market" (June 7, 2001) obtained through the Centaur Communications Ltd. Marketing Week report via Lexis-Nexis. The competition between Pepsi and Coke over the bottled water market are documented regularly in business and financial news reports. The Thomson Corporation's Information Access Company provides reports on market share in the bottled water industry (see, for example, the report of February 12, 2001, for the statistics on Aquafina and Dasani market shares recorded here). A Reuters article, "Coke, Pepsi Ready for Summertime Battle" (May 25, 2001), reviews the ad campaigns of the two cola giants over bottled water. A *New York Times* article, "Just Say No to H2O (Unless It's Coke's Own Brew)" (September 2, 2001) provides an update on Coke's marketing strategies for Dasani. See also "Water, Water Everywhere: Coke, Pepsi Unleash Flood of Ad Muscle" in *The Atlanta Journal and Constitution* (July 12, 2001). The FAO report on bottled water is by M. Latham, "Human Nutrition in the Developing World," Food and Nutrition series no. 29 (Food and Agricultural Organization of the United Nations (FAO)), Chapter 31: "Beverages and Condiments."

For a recent analysis of Coca-Cola, see Mark Pendergrast's *For God, Country and Coca-Cola* (New York: Basic Books, 2000). Several of the background references in the section on Coke Water are taken from Pendergrast's book. Coke's strategy of using mineral packets is outlined in a *Wall Street Journal* report, "The Real Thing: Coke to Peddle Brand of Purified Bottled Water in U.S." (November 3, 1998). The quotes from former CEO Roberto Goizueta are found in Pendergrast's book. Professor Marion Nestle's assessment of the diminished hydration effect of sodas is outlined in an article published by the *Tufts University Health and Nutrition Letter* (July 1998). In India, according to the New Delhi-based National Council for Applied Economic Research, the vast majority of soft drinks are consumed by middle- and low-income people in small towns and villages, not by richer urbanites. The case of Coca-Cola's involvement in Guatemala has been described by Henry J. Frundt in *Refreshing Pauses: Coca-Cola and Human Rights in Guatemala* (New York: Praeger Publishers, 1987). The story about the case being brought against Coca-Cola and bottling franchises in Colombia was initially reported by the *British Broadcasting Corporation* on July 20, 2001, and reprinted in several news wire stories. Ralph Nader's report on the racism charges against Coke regarding its employment practices were printed in the *San Francisco Bay Guardian* (July 29, 2001).

As to available worldwide supplies of fresh water, the most extensive studies to date have been conducted by Igor Shiklomanov at his State Hydrological Institute in St. Petersburg, Russia. Statistics on fresh water sources in the last section of this chapter are taken from Shiklomanov's research as recorded in Peter Gleick's *Water in Crisis: A Guide to the World's Fresh Water Resources* (New York: Oxford University Press, 1993). The full extent of earth's fresh water resources, however, remains unknown, since calculations have not been done on the amount of water in underground permafrost regions or in many marshes and bogs.

CHAPTER 7: GLOBAL NEXUS

The opening story about Oscar Olivera in Cochabamba has been reported in numerous articles, including Jim Shultz's "Bolivia's Water War Victory," published in the *Earth Island Journal* (Autumn 2000). The section on "Corporate Massaging" is based largely on the analysis of the various water industry associations outlined in Chapter 7, "Priming the Pump," in *The Final Frontier* (available at), which draws on information provided by the associations' own websites and written reports. Fay Hansen's article, "Working with International Finance Institutions," published by the National Association of Credit Management (obtained through Lexis-Nexis) provides a useful overview of the role played by the major international financial institutions, particularly the

World Bank's two main divisions (IBRD and IFC) and its regional affiliates (EBRD and ADB). The role of the IMF in "The Push for Water Privatization and Full Cost Recovery" is documented through a set of charts in *News and Notices* (vol. 2, no. 4 (Spring 2001)), published by the Global Challenge Initiative (and also available at). The relationships between governments, the construction industry, and the World Bank are documented in a study entitled *Dams Incorporated: The Record of Twelve European Dam Building Companies*, prepared by The CornerHouse research team in the U.K. and published by the Swedish Society for Nature Conservation (February 2000). Information on the Lesotho Highlands Water Project and the outbreak of cholera can be found in *Watching the World Bank in South Africa*, published by the Alternative Information and Development Centre in Cape Town, S.A.

The section on "World Trade" draws on several background sources. Steven Shrybman's *The World Trade Organization: A Citizens' Guide* (Ottawa: Canadian Centre for Policy Alternatives and James Lorimer, 1999); Lori Wallach and Michelle Sforza's *Whose Trade Organization: Corporate Globalization and the Erosion of Democracy* (Washington, D.C.: Public Citizen, 1999); and Debi Barker and Jerry Mander's *Invisible Government: The World Trade Organization as Global Government for the New Millennium?* (San Francisco: International Forum on Globalization, 1999) provide important background analysis of the WTO. See also Chapter 4 of Maude Barlow and Tony Clarke, *Global Showdown* (Toronto: Stoddart, 2000). Scott Sinclair's seminal work, GATS: *How the World Trade Organization's New "Services" Negotiations Threaten Democracy* (Ottawa: Canadian Centre for Policy Alternatives, 2000) provides valuable insights into the GATS regime. Two other pieces by Steven Shrybman - "A Legal Opinion Concerning Water Exports Controls and Canadian Obligations under NAFTA and the WTO" and "Water and the GATS: An Assessment of the Impact of Services Disciplines on Public Policy and Law Concerning Water" - provide further insights into trade rules affecting water privatization and exports (available through the Council of Canadians). For the section "Regional Blocs" Maude Barlow's *The Free Trade Area of the Americas* (published by the Council of Canadians, March 2001) and Mark Lee's *Inside the Fortress: What's Going on at the FTAA Negotiations* (published by the Canadian Centre for Policy Alternatives, April 2001) both provided useful background analysis.

The section on "Investment Treaties" draws on analysis done to date on both Bilateral Investment Treaties (BITs) and the Multilateral Agreement on Investment (MAI). Michelle Swenarchuk's article, "The MAI and the Environment" in Andrew Jackson and Matthew Sanger's *Dismantling Democracy* (Ottawa: Canadian Centre for Policy Alternatives and James Lorimer, 1998) contains some initial insights about the BITs. Tony Clarke and Maude Barlow's *MAI: The Multilateral Agreement on Investment and the Threat to Canadian Sovereignty* (Toronto: Stoddart, 1997) provides a comprehensive overview of the proposed MAI and the impacts it would have had if this global bill of rights for transnational corporations had been ratified, rather than being rejected, in 1998. It should be noted, however, that moves are being made now to restart negotiations on the MAI through the WTO.

CHAPTER 8: FIGHTBACK

The opening story of the fightback in India's Narmada Valley is based on reports issued by the International Rivers Network in the U.S. See, for example, Patrick McCully's article, "A Stream of Consciousness: The Anti-Dam Movement's Impact on Rivers in the 20th Century," published in *Encompass Magazine* (vol. 4, no. 5, June-July 2000). See also Arundhati Roy, *The Cost of Living* (Toronto: Vintage Canada, 1999). The stories about campaigns aimed at taking back public control of water systems in Cochabamba, Bolivia, and Grenoble, France, are described in an article by Emanuele Lobina entitled "Water Privateers, Out!" published in *Focus*, a magazine of the Public Service International (June 2000). Jim Shultz's "Bolivia's Water War Victory" in the *Earth Island Journal*, as well as the original *Blue Gold* booklet, also contain information on the setting up of SEMAPO in Cochabamba. Monique Bouchard's "Our Fight for Grenoble Public Water Service" (unpublished) also provided helpful insights. The section on "Fighting Privatization" draws on numerous sources. Information on the campaign in South Africa was based on a field trip in May 2001. The "Accra Declaration on the Right to Water" is available from the National CAP on Water via email: . The story of the fightback campaign in Uruguay is based mainly on an interview with Adriana Marquisio, Delegada a la Mesa Coordinadora de Entes Publicos, Federacion de Funcionarios de Obras Sanitarias del Estado, on July 7, 2001. The campaigns waged in the U.S. against the American Water Works Company are outlined in the Canadian Union of Public Employees 2001 annual report on privatization called *Dollars and Democracy*.

The story of the fightback against Perrier and bottled water exports in Wisconsin was reported by *Time* magazine, September 25, 2000. The report on the battle being waged by the Michigan Citizens for Water Conservation against Perrier is based on correspondence and legal submissions to the local government. The consultant firm's report on potential water takings from Aboriginal lands for export was reported by *Canadian Press* on August 23, 2001. The section on "Water Quality" also draws on several sources. In Colombia, the case study cited about the operations of Occidental Petroleum at Caño Limon is reported in *Blood of Our Mother: The U'wa People, Occidental Petroleum and the Colombian Oil Industry* (published by Project Underground, Berkeley California, 1998). Dale L. Watson's unpublished article on "Fresh Water Oil Fields: The Ultimate Bulk Water Export" outlines the campaign emerging in Alberta to resist the heavy use of water to pressure-up oil fields. The *Global Pesticide Campaigner* published by the Pesticide Action Network contains information on campaigns against the use of chemical pesticides in agriculture that destroy water quality ( ). The Water Keeper Alliance's legal challenge against pig factories in the U.S. is outlined in the *Animal Welfare Institute Quarterly* (Winter 2001). And the fight for a public water filtration and sanitation system in Kamloops, British Columbia, is recorded in CUPE's 2001 annual report on privatization, *Dollars and Democracy*.

In the section on "Restoring Watersheds," the reports on Ecotrust and the Haisla people in Kitlope Valley, the Applegate Partnership in Oregon, and the restoration of the Trinity River in Hayfork, California, are outlined in articles published in *Yes! A Journal of Positive Futures* (Fall 1997). The campaign to decommission the dam in order to free up the Kennebec River in Maine was reported in the *Ottawa Citizen*, July 2, 1999. Lori Pottinger's *River Keepers Handbook: A Guide to Protecting Rivers and Catchments in Southern Africa* (published by the International Rivers Network, 1999) describes how "catchment communities" are organized in southern Africa. The section on "Stopping Dams" makes use of Patrick McCully's article "A Stream of Consciousness," cited above, particularly for the story about the resistance to the damming of the Danube in Hungary and the movement for the decommissioning of dams in the U.S. The reports on the struggles waged against the Chixoy Dam in Guatemala and the Pak Mun Dam in Thailand are recorded in *World Rivers Review*, vol. 15, no. 6 (December 2000), published by the International Rivers Network.

CHAPTER 9: THE STANDPOINT

Although we disagreed with most of the contributions in the collection, we read the arguments for water privatization in the 2000 World Bank book, *The Political Economy of Water Pricing Reforms*, published by Oxford University Press and edited by Ariel Dinar. On the other hand, we did make extensive use of arguments put forward by Public Services International and its publications, such as the 2001 *Water in Public Hands*, written by David Hall. *World Water Watch: The Magazine of the Freshwater Environment* (now unfortunately no longer published) was also a source of many of the arguments, for and against privatization. Eric Gutierrez of the Institute for Popular Democracy in the Philippines provided insight in a paper he wrote for a September 1999 conference of the U.K. NGO WaterAid, called "Boiling Point: Water Security in the 21st Century."

Philip Lee of the *Ottawa Citizen* provided a useful series on the solutions to water in late August 2001. Patrick Bond of the University of Witwatersrand in Johannesburg wrote an excellent paper in the summer of 2001 called "Valuing Water beyond 'Just Price It': Costs and Benefits for Basic Human and Environmental Needs." And we wish to give special thanks to colleague Dr. Vandana Shiva of the Research Foundation for Science, Technology and Ecology in New Delhi for many books and publications on the commons. Particularly, we cite her July 1999 paper, "The Politics of Water: Water as Commons or Water as Private Property."

CHAPTER 10: THE WAY FORWARD

The verse at the beginning of this chapter is a Narmada Bachao Andolan struggle song. After hearing it, Patrick McCully had it transcribed into English and included it in his book *Silenced Rivers*. For this chapter, we are most indebted to colleague Riccardo Petrella and the Italian-based global water campaign, the Water Manifesto Project, for his tireless fight to keep water in public hands. In particular, we cite his 2001 book, *The Water Manifesto: Arguments for a World Water Contract*, published by Z Books. Sandra Postel's contribution to the 1996 Worldwatch *State of the World* series is also excellent; it is entitled *Forging a Sustainable Water Strategy*. Peter Gleick, Gary Wolff, Elizabeth Chalecki, and Rachel Reyes of the Pacific Institute have drafted an important report yet to be released, called *The New Economy of Water*. We are grateful to have been given an advance copy.

Friends of the Earth in many countries is deeply involved in working out on-the-ground solutions to the water crisis, and we want to acknowledge their excellent research as well. A 2001 film called *Earth on Edge*, featuring PBS commentator Bill Moyers and produced by the World Resources Institute, provided excellent material on the solutions to South Africa's crisis. We also gratefully referenced the proceeds of an international seminar on small farms held in March 2000, in Katmandu, Nepal. The proceedings were published under the title *Challenges to Farmer Managed Irrigation Systems*.

Surprisingly, a 1999 World Bank-United Nations Development Programme report called *Learning What Works* strongly criticized megaprojects and called for small technologies and community control of water. Finally, we want to acknowledge the campaign in the U.S. Congress to pass legislation that would stop the promotion of water privatization by big financial institutions like the IMF and the World Bank.

Note regarding Metric and Imperial Measurements: Some conversions of metric and imperial measurements in this book are approximate, because of errors due to rounding.

INDEX